Strategies for Successful Science Teaching

Sharon Brendzel

University Press of America,® Inc.
Lanham · Boulder · New York · Toronto · Oxford

4501 Forbes Boulevard
Suite 200
Lanham, Maryland 20706
UPA Acquisitions Department (301) 459-3366

PO Box 317
Oxford
OX2 9RU, UK

Printed in the United States of America
British Library Cataloging in Publication Information Available

Library of Congress Control Number: 2004116236
ISBN 0-7618-3037-5 (paperback : alk. ppr.)

™ The paper used in this publication meets the minimum requirements of American National Standard for Information Sciences—Permanence of Paper for Printed Library Materials, ANSI Z39.48—1984

Dedication

I dedicate this book to my students, our NJ teachers and future teachers whose appreciation keeps me going. They use my ideas and suggestions as the catalyst to provide better science education for their students. This provides me with the incentive to continue my efforts. I am frequently told that if science had been taught to them this way, their lives might have taken a different turn. I expect that this can now be true for their students. The book was compiled at the request of my students and, thus, I dedicate it to them and their students.

Table of Contents

Preface

This book has been written as a text to be used in my science education methods course for elementary school teachers at Kean University. The idea is to provide a simple framework for other instructors to use in their own methodology courses. The basic concepts and strategies are discussed, and examples are presented. Most chapters also have one sample lesson, illustrating the main concept of the chapter. Some chapters have several examples discussed throughout the chapter.

The text, however, has been kept short. The idea is that each instructor will supply additional activities and supplementary materials. Of course, each of us has several favorite activities that we like to use, and some material that we find especially important. The text allows for this flexibility by providing the basics and allowing room for tailoring the course to the individual instructor's approach.

The appendices are detailed, intended to supply a resource to prospective teachers. Each chapter has a reference list and some website citations. A full bibliography is provided at the end.

I hope that you and your students will find this framework as useful as I have.

Dr. Sharon Brendzel
Union, New Jersey
July 9, 2004

Introduction

Good teaching is a complex activity. It requires knowledge of subject matter, theory, and teaching techniques. This knowledge must be blended with creativity, imagination and presentation style. As teachers or pre-service teachers you have been and will continue to develop all of these skills into your own teaching style. This process is on-going and lasts as long as you teach. New ideas and practices need to be blended with the tried and true into the best teaching for your students.

Science teaching, of course, draws on these same components. This book is intended to provide you with some of the skills, tools, and knowledge to create wonderful, meaningful science lessons for your students, who represent the future for all of us.

Some of you are far along on this pathway and only need a few up-to-date ideas and resources to assist you. Others still have a fear of teaching science. It is the aim of this book (and the accompanying coursework) to give you enough information and strategies to feel comfortable with teaching science. It is essential for your students, and for our future as a nation. Fortunately, science activities are fun for elementary children and once you engage your students in the many wonderful opportunities available, both you and your students will enjoy science more.

Chapter 1
Why Science Standards?

"I do and I understand"

"I do and I understand" is part of a statement from a Chinese proverb. It has been quoted frequently by educators, who make the point that learning science should be a participatory activity. This statement is particularly true for young children, who generally respond better to concrete experiences and who develop a *deeper* understanding when they are actively performing the experiment, as compared to being mere observers. Although young children are naturally curious about the world around them, boring classroom presentations of science material can easily stifle this curiosity. This intuitively obvious fact has been borne out through observations in schools, and through quantitative national studies, which have pointed in the direction of engaging children in science activities. Accordingly, the US model for effective teaching currently employs the active participation paradigm discussed here. It should be noted, however, that different approaches used in some other countries can be also successful.[1]

The standards for science teaching, discussed below, emphasize the need to teach children not only through the use of activities but also through the use of critical thinking. Critical thinking is a process involving analysis, synthesis and evaluation.[2] It has been shown that this skill of critical thinking is developed in students by challenging them to think through open-ended activities. Concept

---

1 The Japanese, for example, have been successful in teaching science using a model that employs guided interactive demonstrations, rather than hands-on experiments. This approach to teaching science probably arose from the fact that the Japanese system has very large classes (from 50 - 75 students), and teachers are given more preparation periods to work together to prepare these demonstrations. In this approach, the teacher uses several prepared demonstrations, students observe the demonstrations and the teacher carefully guides them with open-ended questions. An important part of their approach, relative to what has been shown to be necessary for effective teaching, is the use of open-ended questions. The use of open-ended questions is being employed in the US, but the guided demonstrations are replaced with hands-on activities because, for other pedagogical reasons, the US class sizes are smaller, and teachers are given fewer hours for preparation.

2 Bloom's Taxonomy uses this precise phraseology.

development is a natural result of the active engagement of students in the activities.

National Reforms

The national reforms in teaching began in mathematics, with national standards developed by the NCTM (National Council for Teaching of Mathematics). This was followed by the development of national standards for other subject areas, encouraged by other professional organizations. Most states have followed this lead and have developed standards that are in line with the national standards.

The need for a more consistent use of good teaching approaches was obvious to many educators who examined the traditional teaching styles and observed that youngsters were not being motivated to think. Too much emphasis was placed on rote memorization. The poor state of science and mathematics teaching in the US indeed became evident in International Studies, the first of which was conducted in 1988. The results showed that the test scores of US students were significantly lower than the test scores of students from other countries, and provided quantitative and indisputable evidence that supported the views of many educators. We have now completed the fifth round of such tests, and although some progress has been made by US students, they are still not doing as well as it was hoped.

It is in this climate that the national standards were developed. These standards emphasize thinking and problem solving, and to many educators, these standards represent just good common sense. The transformation of the "good common sense" into standards gives all of us a framework against which to check ourselves and, consequently, these standards help us provide the best education for our youngsters.

National Science Standards

The National Science Standards for teaching, in abbreviated form, are presented in Table 1.1. The first column of the table lists the actual standard, and the second column provides a brief explanation of each standard. Each of the standards is discussed below.

Unifying Concepts and Processes

This standard deals with teaching presentation style, and is, therefore, considered to be one of the process standards (in contrast to the content standards). All content teaching should be done with this standard in mind. The notion of this standard is that that the teacher ought to emphasize the umbrella concept being taught, even while teaching the concept details. In doing this, several guidelines for teaching should be kept in mind (Systems, Evidence, Change, Evolution, and Function).

Science as Inquiry

This standard is also a process standard and should be part of the teaching of all content areas. When focusing on inquiry in teaching science, a students' assignment might comprise a problem, and/or an open-ended question with an ultimate challenge to solve the problem. Students should be encouraged to discuss the open-ended questions among themselves and, as a result, form their own hypotheses, design experiments that will prove (or disprove) their hypotheses, perform the experiments they designed, record and analyze the results, and draw conclusions. In the course of discussions, the teacher should point out that in numerous fields the process undertaken by the students is similar to the process used by scientists.

Sometimes, as the teacher, you might wish to point out that in real-life science, one might simply have a set of observations. In such circumstances, the challenge is to derive a scientific principle that explains the observations. To focus on this aspect of scientific investigations, an assignment could be constructed in the same way, using observations as the starting point.

Physical Science, Life Sciences and Earth and Space Sciences

These content standards are broken down into important sub-categories (see Table 1.1). Appropriate goals are listed for each of the sub-categories according to grade level divisions; i.e., K-4, 5-8, and 9-12.

Science and Technology

Technology has re-emerged[3] as an important part of education, especially science education. Curricula should include design activities in which students experience a simulation of developing technology. Curricula should also include information so that students understand the role of technology in today's world.

Science in Personal and Social Perspectives

The Standards include some important issues in science education. These issues are currently considered essential for scientific literacy of an informed citizenry. The topics include Personal Health, Environmental Science, and Technology.

History and Nature of Science

The history of science needs to be included throughout the curriculum in order to satisfy this standard. As the teacher you might want to make students aware that science is a progressive endeavor, and that new scientific discoveries are always building upon previous knowledge. In addition, students need to be aware that scientific principles are derived from trying to fit a theory, or a "scientific law," to observations, and that sometimes such laws are proven wrong, for example, based on more accurate observations, or merely additional observations. Science is, therefore, a living, evolving body of knowledge and teachers should offer the promise that some of the students may, in the future, contribute significantly to mankind by advancing and enlarging this body of knowledge.

---

3 In the 1960s, following the launch of Sputnik, teaching of *pure* science took a more prominent role, sometimes pushing the coupling with technology to the background.

Summary

We need to be aware that, as we teach, it is essential to weave-in various related aspects of science for a complete understanding of the picture. Thus, as we teach a specific topic in earth science, we need to keep the other standards in mind and we need to focus on the concepts of the topic, use inquiry, incorporate technology, consider social issues, and incorporate related history. Of course, the details need to be tailored to the age level and background of the children, but all teachers should be cognizant of the concepts behind the Standards and use them throughout the curriculum. As stated earlier, a summary of the Standards is available in Table 1.1.

Table 1.1 Summary of National Science Standards for Teaching

Standard	Explanation
Unifying Concepts and Processes: Systems — order and organization Evidence — models, and explanations Change — constancy and measurement Evolution and Equilibrium Form and Function	In order to comply with this standard, the focus should be on the "big picture" of the concept being taught, even while teaching the details.
Science as Inquiry: Abilities necessary to do scientific inquiry Understanding about scientific inquiry	In order to comply with this standard, teachers must engage students in actual open–ended investigations.
Physical Science: Properties of Matter Motions and forces Energy	To satisfy this standard, teachers need to cover the identified physical sciences in a manner appropriate to the grade levels taught. Standards are divided into three sections: K-4, 5-8, and 9-12.
Life Science: Living Organisms and Systems Life Cycles Environments, populations and ecosystems	To satisfy this standard, teachers need to cover the identified life sciences in a manner appropriate to the grade levels taught. Standards are divided into three sections: K-4, 5-8, and 9-12.
Earth and Space Science: Earth materials and systems Earth's history and geochemical cycles The solar system and universe	To satisfy this standard, teachers need to cover the identified earth sciences in a manner appropriate to the grade levels taught. Standards are divided into three sections: K-4, 5-8, and 9-12.
Science and Technology Abilities of technological design Understanding science and technology	To satisfy this standard, teachers need to cover the close coupling between science and technology; for example though hands-on technological problem solving assignments that make use of scientific principles.
Science in Personal and Social Perspectives: Personal health Populations and environments Natural hazards and resources Risks and benefits, environmental quality, Human induced hazards Science and technology in society	To satisfy this standard, teachers need to cover current issues that have a science component, so that students learn to critically evaluate such issues, and thus grow up to be more educated citizens.
History and Nature of Science:	To satisfy this standard, teachers need

Science as a human endeavor Nature of Science and knowledge History and historical perspective	to cover the nature of study that leads to scientific advances, the history of those advances, and the role science plays in history and culture.

Correlation with Basic Pedagogical Research

It should be noted that the use of participatory activities that encourage critical thinking fits well not only with teaching common sense and with the national standards but also with accepted standard pedagogical research.

Piaget, for example, suggests that most elementary school students are in the Concrete Stage of development. Thus, teaching elementary school students through the use of hands-on materials positively correlates with Piaget's research. As these students learn something new, they use assimilation and accommodation. According to Piaget, assimilation involves acceptance of a new idea that fits into the existing schema of the student, and accommodation involves changing the student's schema when the new learning does not fit.

Bloom has developed a taxonomy in which he suggests that students learn better when using analysis, synthesis and evaluation. Teaching students by causing them to use critical thinking also fits with Bloom's Taxonomy of Cognitive Objectives.

Bruner has suggested that teaching science in a way that closely resembles the way scientists work yields better learning, allowing the student to discover science. Following the Standards fits well with the ideas expressed by Bruner.

Lastly, a teaching technique that is currently emphasized in various curricula is cooperative learning, which is formalized group learning. Such learning also fits well with the notions of participatory science, discussed above, even with a structure that does not include all of the ideas suggested by the major proponents of the formalized group learning.

Example Activity — Footprint Puzzle

Correlation with Standards

The Footprint Puzzle, presented in Figure 1.1, is an example of a student-engaging activity that satisfies a number of standards, and illustrates the way in which the standards can be incorporated into a classroom lesson.

When this activity is used as part of a university/college course in science education, the activity should be undertaken in small groups. After doing the activity, a discussion should be conducted to assess which of the Standards are satisfied or addressed by the activity, and how the activity might be improved to better satisfy some of the Standards (or improved to touch on additional Standards). The learning approach that is carried out by the use of this activity should be compared to the more traditional teaching approaches.

It is felt that the Footprint Puzzle is a wonderful paper and pencil activity that can be used in lessons of a number of subjects such as earth science lessons, biology lessons, and lessons that integrate a number of subjects (e.g., language arts). The basic idea of this activity has been around since the Earth Science Curriculum Project presented it as a laboratory activity. Many authors and groups have used a Footprint Puzzle similar to the one shown in a variety of activities because the basic idea is rich in possibilities.[4]

---

4 The author's version of this activity was published in *Science Scope* in 1999 ("Prints to Ponder, *Science Scope,* 22,7).

Introduction to Activity

In the school environment, the activity starts by presenting Figure 1.1 to the class with instructions to discuss the footprints found in the figure and to develop a scenario or a story that explains the footprint pattern. Students should be directed to include an explanation for *all* observed patterns, and they should be instructed to assume that what is presented in the provided figure is all of the paleontological "evidence" that is available. That is, the figure is the record as the paleontologists found it. Sometimes, the students are instructed that if they wish to obtain further evidence to help support their scenarios, they may state the type of additional evidence they wish to have (but they may not assume any additional evidence).

Description of Activity

Setting

This activity is usually conducted by dividing the class into groups and having each group do the activity. Working in groups allows for discussions within the group and cooperative learning. A direction can be included for one person to serve as the recorder and other specific assignments can also be made to assist the groups in functioning, as is often done in cooperative learning groups. Experience has shown that this activity involves students in critical thinking, where they challenge each other's interpretations in order to come up with a scenario that the group considers the most likely interpretation of the figure.

After the discussion, each group should write up the group's scenario (providing an opportunity for a language arts lesson). The teacher may insist on a consensus within the group (providing an opportunity in a "civics" lesson), or explicitly allow for disagreement among group members, so that more than one story can be described. In each case, use of supporting evidence should be insisted upon. One student from each group might be chosen to summarize the scenario for his/her group. A class discussion should follow, so that students understand the various possibilities and the supporting evidence presented by the various groups.

Scenario Examples

By way of assistance to the elementary classroom teacher, what follows are scenarios that most commonly have been created.

1. The Fight

Two animals are walking along. At some point, the larger animal sees the smaller one and begins a chase. They meet in the center and fight. Only one animal emerges. This scenario explains the fact that the footprints of the larger animal are farther apart as the animal approaches the center (when the larger animal sees the smaller one). It also explains the confusion of footprints all over the place in the center, and the fact that only one set of footprints is seen emerging from the center. Actually, several different explanations are often offered for the single set of footprints leaving the central part. One explanation is that the larger animal ate the smaller one. Another explanation is that the smaller animal flew away, noting that the footprints are bird-like.

Some elementary school groups report out more details, such as noting that the animal with the emerging footprints has prints that are closer together after the fight than they are coming into the center. This can be explained by the fact that the animal is tired after the fight, is full from eating its prey or is carrying its prey, or simply that it is no longer in a hurry.

Another detail that is often given is a description of the category of the animals. The most common suggestions are that these are bird-like animals. The larger footprints look more like a webbed footed bird, so lizards, other reptiles, and amphibians are also sometimes suggested. Some students think the larger footprints could be an animal with a paw where the claw marks did not get preserved. These are all acceptable explanations, unless detailed examples of footprints are available for better analysis.

2. Mother and Baby

A mother animal is walking along and, when she sees her baby, begins to run. They meet and play in the center, and later walk off, with the mother carrying the baby. This can be accepted because footprints of adult animals do differ from baby animals, and these prints could be in the same family.

3. The Watering Hole or Food Patch

Animals of various sizes are walking to the watering hole, or food source. One animal speeds up upon seeing the water (or food). Not all of the prints are preserved upon emergence. Alternatively, a variation from the fight scenario can be included. For example, a fight ensues at the watering hole, and one animal becomes the prey. Another alternative explanation is that the smaller animal flies off after getting the water or food.

4. Different Time Periods

Since there is nothing in the evidence to indicate the time, these prints could have been made at different times and might have no relationship to each other. This idea can also be combined with the watering hole. If students do not think of this idea, remind them to "break set" and consider the time periods, just to point out another possibility and get them thinking about the need for this additional information.

Lesson Analysis

If this lesson is used as an introduction to fossils and paleontology, students can be encouraged to think about the conditions that result in preservation, the way in which footprints might be preserved in unusual activities, and the methods used to uncover the rock records. Good follow-up questions include the following:

- Which scenario do you think is most supportable from the evidence seen?
- What kinds of activities would provide footprints that could mislead the paleontologist interpreting the footprints?
- What conditions of the ground might alter the interpretation of footprints?

If appropriate, the teacher might want to remind students that footprints can tell us information about the type of animal, what the animal was doing, and the size of the animal. Students can discuss other information yielded by footprints generally.

When this activity is used as a lesson, the Lesson Plan for this activity might state some or all of the following lesson objectives.
The student will be able to:

- Analyze footprints and list specific evidence supporting statements about the footprints.
- List examples of the types of information that can be obtained from footprints.
- Discuss problems that might occur in interpreting records of the past.
- Develop a scenario that adequately explains a footprints record.
- Evaluate various scenarios (presented by the different groups) and select the most logical scenarios from the supporting evidence.

Clearly, this lesson can also be used as an integrated lesson. As a language arts activity, the students concentrate on the story that can be created from the footprints. Students enjoy this springboard and language development is an easy part of the activity.

Another integrated lesson could be to use the activity with a math lesson at the appropriate grade levels. The footprints can be used to create proportions for determining the approximate size of the animals that made the footprint.

As indicated above, the footprint activity can be used at all grade levels, depending upon the introduction and questions asked by the teacher. With very young children, teachers have used the footprints in order to stimulate a discussion about the possible events that took place to create the sequence of footprints. In a college classroom of a general geology course, this activity has been used as a brief introduction to a discussion about the problems faced by paleontologists trying to interpret rock records. Students at all grade levels find the activity motivating and stimulating and it is easy for teachers to use their imagination to enhance the activity. For example, one primary teacher used this activity after a typical early grade hand-mold activity, where the children create a mold of their hands. A middle school teacher laminated the puzzle and actually made it into a jigsaw puzzle, having the students first assemble the picture and then interpret it. Those who have used the activity have reported that this is a fun and successful activity that encourages critical thinking.

Discussion of Correlation with Standards

In this activity the students are developing the ability to use science as inquiry by answering an open-ended question. They are engaged in an active discussion related to the activity. The activity fits most easily into the content area of Earth and Space Science, specifically into the earth's history. The unifying concept is that of fossil preservation. Students are experiencing a portion of the nature of science as they try to formulate hypotheses and discuss the evidence (and lack of it) in understanding the record in the rocks. Technology is not directly involved but can be brought in through the discussion by considering the time period of the particular records. In order to accurately determine the time period, carbon-dating would be needed.

Concluding Remarks

The footprint puzzle is just one example of an open-ended activity that incorporates critical thinking. Throughout the text, several more field-tested examples are given to assist teachers in implementing the Standards and good teaching. It should be remembered, however, that these are but examples of the kinds of activities teachers ought to use in their science teaching.

Figure 1.1 The Footprint Puzzle

Selected Sample Websites

http://www.njpep.rutgers.edu
(The text of the *National Science Standards* on line)
(NJ Science Standards and information about ESPA and GEPA).

(National Academy Press - texts and standards-related material)
To obtain the science standard websites of the different states it is recommended that an Internet search be carried out (using any conventional search engine, such as Yahoo) for the topic "science standards." This leads to a list of science standards by state. Several states actually include the standards; others have documents relating to them. For New Jersey, for example, the site is listed above.

References

American Association for the Advancement of Science. (1993). *Benchmarks for Science Literacy: Science for All Americans.* Washington, D. C.: AAAS.

American Association for the Advancement of Science. (Spring, 1995). "Common Ground: Benchmarks and National Standards." *2061 Today.* Washington, D. C.: AAAS.

Bloom, B. S. (1956). *Taxonomy of Educational Objectives. Handbook I. Cognitive Domain.* New York: McKay.

Brendzel, Sharon (1999). "Prints to Ponder" *Science Scope*, 22,7.

Bruner, J.S. (1963). *The Process of Education.* New York: Vintage Books.

Graika, Tom. (1989). "Minds-on, Hands-on Science" *Science Scope,* (March).
The International Assessment of Educational Progress. (1992). *Learning Science.* Princeton, New Jersey: Educational Testing Service.

Johnson,D.W. and R. Johnson. (1975). *Learning Together and Alone.* Englewood Cliffs, NJ: Prentice-Hall.

Lapointe, Archie E. et al. (1989). *A World of Differences: An International Assessment of Mathematics and Science.* Princeton, New Jersey: Educational Testing Service.

National Research Council. (1990). *Fulfilling the promise: Biology Education in the Nation's Schools.* Washington, D.C.: National Academy Press.

National Science Education Standards. (1996). Washington, D. C.: National Academy of Science.

Piaget, Jean and Barbel Inhelder (1969). *The Psychology of the Child.* New York: Basic Books.

Slavin, R. (1983). *Cooperative Learning*. New York: Longman.
TIMSS (1997). *Third International Mathematics and Science Study*. Philadelphia, PA:

Mid-Atlantic Eisenhower Consortium for Mathematics and Science Education.
"What Research Says" (9/82), *Science and Children.*

Chapter 2
Building the Basics, Process Skills

Introduction to Process Skills

Science content should be taught through participatory activities involving the students in classroom investigations. In performing these investigations and participating in activities, students necessarily use various skills that relate to the processes involved. These skills are typically referred to as process skills, and educators need to teach in a manner that enables students to know how to use these skills.

Often the skills are divided into the "Basic Process Skills," such as observing, measuring, classifying and recording data, and the more complex skills (called the "Integrated Skills"). These integrated skills utilize the basic skills, such as hypothesizing and interpreting data. Table 2.1 is a modified list of process skills presented by the AAAS (American Association for the Advancement of Science), but it should be understood that the set of process skills recognized by different professional organizations, or authors, does vary somewhat. Thus, the AAAS does not include communication as one of the process skills but this important skill is listed by several other sources. An example for each of the skills is included in table 2.1, and each of the skills is discussed following the table, including one or more simple illustrative activities for teaching specific skills.

Although investigations typically require numerous skills, for teaching purposes, it is sometimes necessary to focus on a particular skill by using a simple activity that emphasizes the specific skill. For example, the skill of observing is clearly used in every scientific investigation, but it is important for students to develop this skill in order to become careful observers and, therefore, one or more lessons may need to be included to focus on this skill.

Table 2.1 The Process Skills

Process Skill	**Definition**	**Activity**
Basic Skills		
Observing	Using all five senses	Identifying a peanut Observation boxes
Classifying	Grouping objects and events according to their properties	Sorting buttons
Space/time relations	Visualizing and manipulating objects	Growth of a plant
Using numbers	Using quantitative relationship	Growth of a plant
Measuring	Expressing the amount of an object in quantitative terms	Growth of a plant Metric Olympics
Inferring	Giving an explanation for a particular object or event	Footprint puzzle Magnetic, non magnetic aluminum shapes
Predicting	Forecasting a future occurrence	Growth of a plant Magnetic objects
Integrated Skills		
Defining operationally	Concrete descriptions	Friction
Formulating models	Constructing images, objects, or mathematical formulas	Lung model
Controlling variables	Manipulating and controlling properties in order to make comparisons	Growth of a plant
Interpreting data	Arriving at explanations, inferences, or hypotheses from data	Growth of a plant
Hypothesizing	Stating a tentative generalization of observations or inferences	Growth of a plant
Experimenting	Testing a hypothesis	Growth of a plant

The Basic Process Skills with Examples

Observing (Identifying a Peanut activity)

A good example of a simple activity that can be used to focus students on the process of observing is one where students are challenged to observe a specific peanut in order to later identify it. In this activity, each group of students is given a peanut from a pile of peanuts, with the instruction to observe the peanut carefully so that it can be picked out from the pile of peanuts after it is returned to the pile. The teacher should not mark the peanut in any way, and the students should also be instructed not to mark their peanut in any way. After a given period of observations, the peanuts of the student groups are returned to the pile, and students are asked to pick out their peanut from the pile. It is helpful, of course, if the pile of peanuts is manageable in size, for example, 25-30 peanuts, and at least a few of the peanuts have distinguishing marks (allowing these peanuts to be easily identified, or eliminated, by the group, as appropriate).

Variations on this activity can be easily constructed by using different objects. Near Halloween, for instance, small pumpkins can be used. Even a single object can be used, such as a single large pumpkin, with the activity concentrating on observations by the class and making a list of distinguishing characteristics. In such a case, students might be instructed to list ten or more observations that might distinguish the one pumpkin from other pumpkins. This instruction might be needed because, otherwise, students would probably describe the pumpkin with only three or four characteristics.[1]

Sample Directions for Students: Carefully observe the object given you, recording your observations. The goal is for you to be able to recognize your specific object from the group of similar objects.

Classifying (Sorting Buttons activity)

One activity that is commonly used and is very effective involves classifying buttons. This works well because buttons come in many sizes, textures, colors, materials etc., and there are several appropriate ways to classify buttons. The activity typically stimulates a good discussion about procedures, specifically the criteria for classifying. Students quickly realize that the criteria used for classifying affects the classification results and possibly also affects the sorting method that is selected. For the purpose of understanding the concepts used when classifying, a teacher might choose to accept any reasonable method that a student group uses. Alternatively, the teacher might specify the classifying criteria.

Sample Directions for Students: Separate the objects given to you into groups based upon reasonable categories of your choice. How did you decide which categories to use?

Space/Time Relations (Plant Growth Activity)

Most activities naturally use space or time. This notion is formalized as a basic skill as a reminder to teachers to use actual experiments (using physical

---

1 For an additional observation activity refer to "Observation Boxes" in the activity section of this chapter.

objects) rather than only using paper-and-pencil activities. An experiment on plant germination and growth involves both space and time. In this activity, plants are grown over a period of time and the growth is measured. Students test the effect of various factors on plant growth (for example, water, sunlight, nutrients, etc.). Since a growing plant requires more space as it grows, the need for the increased space should be noted in the course of the activity.

Teachers ought to keep in mind that time considerations occur mostly when rates are measured (e.g., growth rates, evaporation rates, speed, etc.).

Using Numbers (Plant Growth activity)

Using numbers to quantify things is part of everyday human life. We deal with how much money to take with us when we go out of the house, with how many eggs we wish to buy, with how many pieces of lumber we need to build a deck, how many violins we wish to include in an orchestra, etc. There is practically no human endeavor that does not use numbers. Scientific investigations are certainly no exception, except perhaps that science uses numbers more than most fields. Therefore, students should get used to the idea that science is quantitative, and when teaching science, quantitative data should be included wherever possible, rather than avoided.

For example, the plant growth activity described above can be made to include numbers, by directing students to measure and record the plant's growth every day (or some other time interval). Thus, time is also incorporated. The incorporation of time can take place in any other activity that takes time to develop, for example, water evaporation from a beaker.

Some experiments also involve the use of numbers to derive other numbers through mathematical procedures, for example, the mathematical procedure for computing percentages, or an area of a square; and these procedures (formulas) should be applied in age-appropriate settings. In other words, students should become aware of, and comfortable with, the use of numbers to derive other numbers. An example of an appropriate activity might be to determine the average rate of growth at the end of the above-described plant growth activity. At higher grade levels, the teacher can also hold discussions about the fact that at different time intervals the rates of growth were different, and on the meaning of an *average* rate of growth.

Measuring (Metric Olympics activity)

Practice in measuring can be part of many experiments and students should gain experience with all types of measurements, such as volume, distance, temperature, weight, and time. The teacher should have the students use different measurement systems as well. For example, in connection with linear, volume and weight measurements, the metric as well as the English system should be used; and in connection with temperature, the Centigrade as well as the Fahrenheit system should be used.

When focusing on measurements, particularly in lower grades, the emphasis should not be on conversions of measurements from a value expressed in one system to a value expressed in another system. Rather, the focus should be to learn to measure accurately, and to get a sense for the obtained numbers. For example, 25 °C is quite warm but 25 °F is quite cold, and 100 °C is very hot (boiling water) but 100 °F is quite a bit less hot. On the other hand, when focusing on the tie-in between physical measurements and mathematics, particularly in at higher grade levels, teaching how to convert from one measurement system to another (i.e., use of the formula/procedure) is quite appropriate.

There are also several activities that are almost like games, making the task more fun. A good example is the AIMS "Metric Olympics"[2]. In this activity students use simple objects as a javelin (straw), shot put (cotton ball) and discus (paper plate). The object is thrown, and the distance is measured, in the metric system. Students make predictions and compare the predictions to the actual measurements. Another part of this activity is the "Marble Grab," where students are asked to estimate how many marbles they can grab and retrieve from a container with one hand. Although the original activity includes

Inferring (Magnetism activity)

The process skill of inferring is part of most investigations. It is an important critical-thinking skill, where a student reaches conclusions based on indirect evidence. Of course, children need experience in using this skill in order to help them develop the ability to infer.

One simple activity that can be used is an activity in which students need to make inferences as to whether objects are magnetic. In this activity, students are given two objects, a piece of aluminum shaped into a rectangle and another piece of aluminum shaped into a ball. They are also given a magnet to use for testing. The aluminum piece shaped as a ball contains iron filings or small steel pellets, but these are not visible. Students are asked to make comparisons of the two objects using the magnet as one of the observation tools, and then to infer the probable cause of the observed differences. Once the inference is made, students may test the objects in any safe way they think of in order to explain the difference.

A number of students will use the provided magnet and will infer that the round object contains some magnetic material. Although other students will focus only on shape, a simple test will reveal that this is not the difference. Thus, re-shaping the rectangle into a ball does not make it magnetic.

Sample Directions for Students: Examine the objects given to you. How are they different? Why is only one object magnetic? What is your explanation of the difference? Consider your inferences, and then test your ideas.

Predicting (Magnetic vs. non-magnetic activity)

In most experiments students should be encouraged to predict the results of the experiment, as a means to involve them in the activity. It also gets them used to the idea of thinking the experiment through.

A simple activity for prediction involves the use of a magnet and several objects that can be classified as magnetic or non-magnetic. Obvious examples of steel items and non-magnetic objects of paper, cork, plastic etc. should be included. Metal objects that are not magnetic (aluminum and copper, for example)

---

2 The activity is discussed in more detail in Chapter 7.

should be included, so that students become aware of the fact that not all metal objects are magnetic.

Sample Directions for Students: Examine the objects given to you. Decide which objects you expect to be attracted by a magnet and which objects you do not expect to be attracted by a magnet. Write your predictions down.

Now experiment, to test your predictions.

Communicating (Describing an object activity)

This skill should be a part of all activities when students are working in their groups. That is, the students should talk with one another in their groups and all of the students should participate in the discussion, rather than allowing one student to do the work while the other students remain passive, silent, observers. If a teacher finds that students need assistance in communicating well, a simple activity can be to have students describe a simple everyday object with sufficient detail for other students to guess what the object is. Students can select the object themselves and try to describe it for a partner without actually naming the object. The partner needs to guess what the object is. To make the activity more interesting, teachers may impose some restrictions. For example, the object can be described in the format of the game "Twenty Questions," or the object can be described by setting a specific number of descriptive words, or the object can be described by a drawing only. Any of these requirements can also have a time limitation. Students will often be surprised to see how difficult it can be with some objects.

Integrated skills

As stated earlier, and summarized in table 2.1, the integrated skills are the more complex skills that encompass use of some of the basic skills. Integrated skills are an integral part of performing all scientific investigations. In other words, a scientific investigation always incorporates the use of at least several of these skills. As students perform an experiment in the course of learning some substantive subject matter and use one or more of these skills, the teacher needs to verify that the students can use the skills appropriately. Should the teacher determine that the students' skills are in need of improvement, the teacher might wish to focus on improving some or all of the skills through appropriate activities, as illustrated below.

Controlling Variables (Plant Growth activity)

An experiment is an activity undertaken to learn more about a particular situation by taking an action and observing the consequent results. For example, an experiment that consists of pouring water into a beaker containing a sugar cube may be the *action*, and observing that the sugar cube dissolves is the *result.* Several factors affect the experiment's results, and these factors are called variables. For example, the size of the sugar cube, the amount of water poured in, and the temperature of the water affect the dissolving process, and these are some of the variables in the experiment. Obviously, if an experiment is conducted to determine the effect of an experiment's manipulated variable on the observed results (for example, the effect of water temperature on dissolving sugar) one must keep all other variables from changing. Otherwise, one would not be able to draw a conclusion as to the cause of the observed change. That is, one would not know whether the observed experiment results are due to the manipulated variable (often called the "variable") or due to changes in the other variables. Therefore, standard practice includes controlling all variables except the one being tested and, for any given experiment, the set-up of the controlled variables is referred to as the "control".

The activity on plant growth can serve as a useful tool in helping students to understand the use of controls and variables. In the experiment on plant growth, only one variable should be changed, and all other conditions should be as similar as possible. Students should choose the variable that they wish to test. If they are testing plant growth as result of different amounts of nutrients, the same types of plants should be used and all other variables are kept the same except for the amount of the specific nutrient to be tested (the control receives no nutrient supplement). Students should be aware that they need to compare the plant with nutrients (the variable) to the plant without nutrients (set up as the control). If the students observe that the plant with nutrient additives grows faster and looks healthier, they can attribute the result to the nutrient used.

A more thorough and complex experiment on the same topic would be to use several different quantities of nutrient additives and chart the best results, determining the maximum amount of nutrient additives producing positive results. In a real world situation, several similar tests should be used to help support the veracity of the results. Typically, in the classroom, comparisons made between the different groups conducting the experiment are considered sufficient.

Hypothesizing (Plant Growth activity)

Forming a hypothesis means making a statement about a concept that one believes to be true. Such a statement might postulate an explanation of why something is happening (e.g., something we call friction causes moving objects to stop) or might try to identify a factor that affirmatively affects something (e.g., aspirin does relieve headaches). The hypothesis, therefore, is an "educated guess," with the accent on "educated." Accordingly, an hypothesis should be based on information and/or research done before the experiment takes place, and the purpose of the experiment is to provide evidence that supports the hypothesis, or discredits it. In the real world, scientists often do a great deal of research before forming an hypothesis and, sometimes the hypothesis comes from observations and conclusions of a previous experiment. For students, the hypothesis is usually based upon previous learning and common sense experience.

The above-described plant growth activity may be used effectively to focus students on the skill of forming hypotheses. To provide such a focus, students may be asked to construct an experiment with nutrients (e.g., chemical fertilizers) being the variable, and other factors (watering, sun, etc.) being the control. Students are likely to hypothesize that nutrients would probably help plants to grow because they have seen plants and lawns being fertilized. Some students may even be aware that excessive amounts of nutrients are counter- productive. If several different amounts of nutrients were to be used, the hypothesis would include the idea that moderate amounts of nutrients are beneficial, while excessive amounts are harmful.

Experimenting (Plant Growth activity)

The experiment is the actual test of the hypothesis. That is, the aim of an experiment is to determine whether the hypothesis is correct. If the hypothesis is proven incorrect, the gathered evidence often leads to a new hypothesis (that is a better "educated guess"). Conducting the experiment properly requires the use of other skills. First one needs to decide what notion to investigate, then one needs to form a hypothesis, decide on which factors are included in the control, and which factor is the manipulated variable. Carrying out the experiment also involves collecting the appropriate materials, and determining the specific procedural steps. That is, one needs to determine how the manipulated variable will be changed in the course of the experiment, and how the effects of these changes are to be recorded. For example, students should prepare a chart, or a table, for recording of the data. Once the data is recorded, it needs to be interpreted, which may include creating a graph based on the recorded data, or performing calculations using the data. The interpretation will tend to either support the hypothesis or discredit it. The hypothesis cannot actually be proven or disproved with a single experiment.

In the experiment on plant growth students should first discuss and identify all factors that might affect plant growth. This may be, for instance, type of plant, type of soil, size of container, sunlight conditions, watering schedule, temperature, and added nutrients. A teacher might allow different groups to choose a different factor as the variable, with the other factors forming the control, or might specify the variable. With respect to the factors in the control group, they need to be fixed (for example, where the plants are placed, or what the water schedule will be). With respect to the variable, one needs to collect the plants, potting soil, containers, watering devices, and locate a reasonable spot to set up the plants. The variable to be tested needs to be selected. When nutrients are chosen as the variable, one needs to choose the specific nutrient, and the amounts to be used. Then, charts need to be set up so that records can be kept of plant growth and condition. During the experiment, the plants need to be watered (part of the normal condition) the same amount each day, and changes in plant growth, such as size and general health, need to be recorded in regular intervals (usually every few days).

Interpreting Data (Plant Growth activity)

Interpreting data involves studying the information recorded in the course of conducting the experiment. In order to better understand the data, sometimes it is helpful to plot it in a graph, or perform some calculations on the data. Thus, in the plant growth experiment, for example, students should measure the plant growth and record the general health of the plants. If the plant grows taller and appears healthier with a specific amount of nutrients, students should be able to conclude that the nutrient does appear to be useful to plant growth. If different amounts of nutrient are used, and students collect results that indicate poor growth beyond a certain level of nutrient addition, students should be able to conclude that a range of nutrient additions promotes plant growth but that beyond a certain range the addition of more nutrients is destructive.

It should be noted that "data" refers to quantitative measurements and recorded numbers (for example, the height of the plants in the above experiment). This quantitative kind of information provides the most valuable experimental evidence. However, interpretations can also be made from qualitative observations (for example, records of the plant health in the above experiment). These qualitative interpretations should also be included in classroom experiments. For students learning about science (as opposed to scientists in a research laboratory), the qualitative results are important because these results may be more concrete than the collected data. In the example chosen, both the quantitative and qualitative observations are concrete, but in many investigations the quantitative data require complex interpretations. For teaching purposes, it is important to include quantitative data that is simple to interpret and is concrete. For example, in the case of plant growth, some plants will be visibly taller than others.

Defining Operationally (Friction activity)

This skill, which is included in table 2.1, relates to the ability of students to derive concepts, and effectively create definitions, as a result of observations or data obtained from experiments. Friction is a good example of an activity where an operational definition can easily be a part of the experiment. In an illustrative friction activity, students are given a block of wood and a collection of surfaces (carpet, sand paper, wood, waxed paper, plastic etc.). The students are then asked the questions "On what surfaces does the wood block slide more easily? Why?" Then they are encouraged to perform whatever experiments they wish in order to answer the question. Students discuss the assignment among themselves, design one or more experiments, form an hypothesis that answers the activity's question, perform their experiments, make the corresponding observations, and determine whether the experiment results support their hypothesis. Typically, students reach a conclusion involving the roughness, or smoothness, of the surfaces.

In the class discussion that follows the students' experiments and their answers to the question posed by the activity, students normally have little difficulty in generalizing that the roughness of a surface is the major impediment to moving an object across a surface. The teacher can then conduct a discussion about the activity, asking questions about the surfaces and guiding the discussion to the conclusion that the most important impediments to the sliding are the roughness of the surfaces (the sliding surface of the wood block and the surface on which the wood block is sliding) and the size of the surfaces that are sliding against each other. The sliding of one surface over another surface is a rubbing together of surfaces, and once the students understand the concept that rubbing of surfaces against each other is what hinders sliding, the teacher introduces the word, in this case, friction.[3]

---

3 At times, a student might ask where is the friction (rubbing of surfaces) when wheels are used – for example in a toy car. The teacher might point out that friction associated with rolling an object on wheels is, indeed, much lower than sliding the same object

The experiments that the students may select can vary considerably. Students may simply push the wood block and sense the amount of force that they need to exert; students may give the wood block a push and observe how far the wood block moves before it stops; students may place the surfaces on an incline, allow the wood block to slide down from the top of the incline, and measure the time it takes before the wood block reaches the bottom of the incline; etc.

When possible, definitions should be developed this way, rather than given prior to the activity.[4] Of course, some definitions are essential in order to conduct the experiment. Even though most definitions are not needed before an experiment is conducted, the teaching pattern in the past generally has been to give the definition up front, rather than allowing the children to develop whatever possible operationally from the experiment. It is believed that, when preparing the lesson plan, the teacher should decide which terms, if any, are essential to be given out prior to the activity, and whatever else needs to be learned should be developed in the course of the activity and the discussions that follow.

Formulating Models

In the area of science, a model typically refers to a physical representation of something else, often in a simplified way or in a modified scale. A simplified model of a complex system permits students to focus on the important aspects of the system that is being studied, and thus better understand it. One example is the standard model of the earth's structure, from the crust to the core. Similarly, a model where the scale of the system being studied is modified to make it visible and manageable in size also permits students to better understand it. Classic examples of the latter include models of the atom, and models of the solar system.

It is well recognized that models are very useful, particularly in teaching science, and therefore, a teacher can find many commercially available models that can be acquired and used. However, in many circumstances, it is much more effective and educational for the models to be constructed by the students (or the teacher, if the construction is too difficult, complicated, or hazardous) from everyday components. The teacher should also realize that drawings are also, effectively, models.

At higher-grade levels, the teacher might point out that even equations are effectively models. They may be considered to be models because they can predict some aspect of the behavior of a system represented by the equation, and often the equation provides a good insight. For example, the equation which states that speed of a falling object is equal to the gravity pull times the time of the fall, squared[5] is effectively a model that can give students a good insight into notions of acceleration.

(without wheels). The rubbing of surfaces still exists between a wheel and the surface it rolls on, and also between the wheel and the axis around which it turns.

4 Other definitions that can be developed operationally, for example, are density, acceleration, saturation point, diffusion, etc.

5 Often expressed as $s=gt^2$, where s is the speed (for example feet per second), t is time (seconds), and g is the gravitational pull (32 feet per second squared).

Every science area has many useful models that children can easily build, allowing for visualization and a better understanding of the concept. Examples include working body parts such as the lung, cells etc., earth models and earth system models, atomic structure models, simple machines etc.

Example Activity — Observation Boxes

Correlation with Process Skills

Process Skills are used in all investigations and, usually, several process skills are used in the course of an investigation. Throughout this chapter, examples of activities have been given which will help students to understand and use process skills. What follows is a detailed lesson plan for an activity that focuses on the process skills of observation and interpretation (inference). It should be noted that this lesson can also be used as part of a content unit (for example, on senses). In the university/college classroom, when this activity is undertaken, students should explore the various classroom uses to which this activity lends itself. They should discuss problems that might occur, questions that can be formulated, concepts that can be emphasized and other examples that might be used.

Introduction to Activity

This activity, which was adapted from the ESCP curriculum[6] (1967), consists of a series of shoeboxes that contain different items. In connection with each box, students are directed to use a specific sense or technique as a means for observation. From these observations, an interpretation is made as to the actual contents of the box. The senses used include sight, touch, sound, indirect touch and all senses (optional box). Thus, the general objectives include teaching the students to make good observations, and helping the students to understand the importance of using as many senses as possible when making observations.

The directions given on the following pages are directions that a teacher might use in the elementary school classroom (or in the university/college classroom as a simulation). The first direction is a general direction that relates to all of the boxes, while the subsequent directions are the directions that are associated with the corresponding boxes. The directions also provide explanations as to how the teacher can prepare these boxes for the students, including suggestions for specific items that might be included in each box. It should be noted that this simple activity can be used at several grade levels and with many variations, and some of these variations are described in the "Lesson Analysis" section. A student activity sheet is provided in Figure 2.2

Description of the Activity (Teacher Portion indented)

General Directions (for students):

Observations - For each box, describe your observations using the sense indicated in the instructions attached to the box.

---

6 Earth Science Curriculum Project, (1968) *Investigating the Earth.* Boston: Houghton Mifflin.

Interpretation - Based on your observations, make an interpretation (inference) as to the nature of the object or objects insides.

Specific Directions (These directions are for teachers and students, labeled appropriately):

Box 1—Sight

Teacher Portion: Set-Up — Place a number of common objects in the box, allowing them to be clearly visible. The best choices include objects that can be identified as a group, but can be identified in more detail with appropriate information. An example might include a shell and a rock.

Comments — Students often find that they want to use various senses in their detailed observations, but the restriction to use sight only makes them aware that, although the sense of sight is the most important sense used for observation, it is easier to be able to use all of the senses in completing an investigation. Students also become aware that additional research may be helpful for proper identification of what has been observed, such as identifying the types rocks or shells that are found in the box.

Student Directions: Examine these objects by using your sense of sight and describe the characteristics from this viewpoint. Do not touch the objects.

Box 2—Sound

Teacher Portion: Set-Up — Place a number of objects in a box, cover it, and place a rubber band around the box so that the cover will not fall off on its own. It is best to include small identifiable objects with a variety of characteristics. Buttons are a good possibility because they come in various sizes, shapes and materials. It is also a good idea to include another object whose sound is easily muffled by the majority of objects in the box. A spool of thread works well for this choice.

Comments — Students become aware that many objects have similar sounds, that louder sounds may muffle softer sounds and that some groups of objects have many sounds.

Student Directions: Lift the box off the desk and gently shake it, but do not open the box. Describe the properties of the objects inside.

Box 3—Touch

Teacher Portion: Set-Up — Place objects in a box with one end open but covered with a flap. Familiar objects with different textures are best, such as "bean bags" stuffed with various materials of different kinds, sand, cotton, or wax paper.

Comment — Students learn that a large amount of information can be obtained simply by touch. Some students are able to identify the specific beans; some students correctly count the objects. Others are less careful in their observations, but these students should become aware that more careful observations could be made.

Student Directions: Put your hand inside the box and feel the object or objects. Do not open the box to make your observations, and do not move the box.

Box 4—Indirect touch

Teacher Portion: Set-Up — Place objects in a box whose top has several holes punched out. Students are provided with a stick to poke inside the box. (Chop sticks work very well for this activity.) The best objects to use are small stuffed animals. They have different shapes and include different textures.

Comment — This box is the most exciting one for youngsters. Only a few of them guess the object correctly. Some youngsters hit the hard eye, but most do not. This activity leads into a discussion of the difficulties scientists have when data is indirect and actual materials cannot be seen. Sonar, radar, and core samples exemplify this type of data.

Student Directions: Use the stick to poke through several holes in the box. Do not lift or open the box to make your observations. Describe the characteristics of the objects.

Box 5 – A number of senses

Teacher Portion: Set-Up — At least some of the objects that are placed in the box should be ones that are difficult to identify even when using all of your senses. Good examples include silly putty or "goop" (a mixture of starch and water that is solid under pressure and liquid when the pressure is removed). Both of these materials do not follow the "normal" pattern of observations. Even though all of the senses are available these objects are difficult to describe.

Comments — Students learn that some materials are difficult to categorize. They behave in unexpected ways, and careful observations and interpretations are needed. Students typically note that using as many senses as possible makes the observations easier to accomplish.

Student *Direction:* You may touch, smell, lift or examine the material in this box in any way that will help you determine its properties.

Figure 2.1 – Observation Boxes

General Directions:

Observations — For each box, describe your observations using the sense indicated in the instructions attached to the box.
Interpretation - Based on your observations, make an interpretation (inference) as to the nature of the object or objects inside.

Specific Directions:

Box 1 — Sight

Directions — Examine these objects by using your sense of sight and describe the characteristics from this viewpoint. Do not touch the objects.

Box 2 — Sound

Directions — Lift and gently shake the box. Do not open the box. Describe the properties of the objects inside.

Box 3 — Touch

Directions — Put your hand inside the box and feel the object or objects. Do not open the box to make your observations. Do not lift the box.

Box 4 — Indirect touch

Directions — Use the stick to poke through several holes in the box. Do not lift or open the box to make your observations. Describe the characteristics of the objects.

Box 5 — All senses

Directions — You may touch, smell, lift or examine the material in this box in any way that will help you determine its properties.

Lesson Analysis

This activity can be used in almost all grades. It was originally planned for a ninth grade class, but the directions can be simplified, if felt necessary, for the lower grades and students can work in groups and report orally. The activity has been successfully used with several grade levels. It has even worked well with students as low as grade one where it can be used as part of a unit on the senses. At the upper levels, more discussion can take place about the relationship of the set-up to scientific endeavors and the difficulties encountered when information is incomplete.

Discussion of Correlation with Process Skills

The activity on "Observation Boxes" is an example of an activity that focuses on observation and interpretation or inference. The teacher should be aware that students sometimes have difficulty in distinguishing observations – which relate to unbiased reporting of what the senses perceive, from the interpretations – which adds judgments and conclusions to the observations. A classroom discussion should focus on this difference. Similar types of activities should be used to assist students in the use of all process skills.

Concluding Remarks

It should be readily apparent that the activity on "Observation Boxes" employs several of the process skills including observation, prediction and, interpretation. It can also be easily adapted, using parts of the activity to emphasize other process skills such as classification or communication. Throughout the chapter other activities were also presented and discussed in terms of teaching process skills. It should be clear that these process skills are an important part of all investigations and activities. If teachers observe that students need a review of process skills, such a review should be conducted, and students should continue to use process skills and remain aware of their importance.

Selected Sample Websites

http://www.mcel.org/whelmers

Lesson plans organized to correlate with the *National Science Standards*. Each lesson lists the process skills used, the standards and a full lesson plan for conducting the activity. Interesting though questions are also provided. Additional materials can be purchased in book form.

http://www.aaas.org/

American Association for the Advancement of Science.

This organization has many on-line services. The Science Education section leads to information about *Benchmarks* and other current materials.

http://www.enc.org/

Eisenhower National Clearinghouse

This organization provides educational publications, free materials, extensive background material for teachers, lesson plans and activities.

http://www.utm.edu/departments/ed/cece/SAMK8.shtml
The Science Activities Manual (Sam) is correlated with the national standards and organized by grade level and science topic. It was prepared to support the Tennessee Science Curriculum Framework. The emphasis is on hands-on activities.

References

AIMS (Activities for Integrating Mathematics and Science). AIMS Education Foundation, PO Box 8120, Fresno, CA 93747-8120.

Brendzel, Sharon. (1992)."Science Education at a Crossroads," *Educator's Clipboard.* 4, 2, Fall.

Earth Science Curriculum Project. (1967). *Investigating the Earth.* Houghton Mifflin Company: Boston, Massachusetts.

Chapter 3
Focus on Inquiry

Introduction

> The *Standards* call for far more than "science as process," in which students learn skills as observing, inferring, and experimenting. *Inquiry* is central to science learning. When engaging in inquiry, students describe objects and events, ask questions, construct explanations, test those explanations against the current scientific knowledge, and communicate their ideas to others. They identify their assumptions, use critical and logical thinking, and consider alternative explorations.[1] (emphasis supplied)

The above quote indicates the importance of the inquiry approach in teaching science. Specifically, the quote states that science teaching needs to go beyond the process skills, incorporating the inquiry approach. In this approach students are fully involved in the scientific process. As they engage in inquiry investigations, they are using process skills, scientific procedures, critical thinking, problem solving and they are enjoying the activity. Inquiry activities can be adapted for all age levels with appropriate activities. Most inquiry activities start with a good open-ended question for students to answer. The question needs to be one that students can discuss and work towards solving.

Thus, we see that the term inquiry reflects an attitude and an approach to the teaching of science. As already stated, *The National Science Education Standards* encourage this approach, stating that teachers should "focus and support inquiries as they interact with their students".[2] For students, another benefit of this approach is that it results in better retention and understanding of the concepts.

Inquiry in the science classroom encompasses a range of activities. The breadth of this range should, of course, be related to the grade level of the students. Some activities provide for observation, data collection and analysis. Other activities involve students in the design of open-ended activities related to a teacher-posed question, or questions that arise from classroom discussions. When students don't formulate their own questions, the teacher poses open-ended questions or suggests open-ended activities.

An open-ended activity is one with an open-ended question which basically has more than one correct answer and which, by its nature, encourages divergent

---

1 National Science Education Standards, p. 5
2 National Science Education Standards, p.33

thinking. Open-ended activities are particularly effective in encouraging critical thinking and providing a basis for connections to other subjects. A fairly comprehensive discussion of student inquiry in the science classroom is found in *The National Science Education Standards* [3]. These concepts are also found in *Benchmarks for Science Literacy* (American Association for the Advancement of Science) and other national and state standard documents. While it is important to note that these concepts are currently stressed by National and State organizations, the importance of the inquiry approach actually rests on educational research, scientific procedures, and the behavior of students in the classroom.

Implementing Inquiry

Most textbook activities today offer many good ideas for teaching activities, and they are easy to use. However, the activities are usually in the form of a specific set of directions, akin to a recipe. A few new programs have been developed to emphasize inquiry, but most activities are not inquiry-based. It is therefore incumbent on teachers to develop such activities, ant it is important that administrators be prepared to assist teachers wherever possible. Many colleges and universities offer workshops on inquiry methods, and some of the new materials that have been developed in the workshops can go hand-in-hand with regular textbooks. Also, some products which include inquiry-based teaching materials can be purchased from various entities. Most science education texts include descriptions of several of these programs.[4] In this text, Appendix A includes a list of some of the most popular materials, Appendix B includes appropriate activity books, and Chapter 11, "Resources" includes a discussion of some examples.

One of the most effective approaches is to adapt the activities on hand. Often, it is easy to open up activities. It can be as simple as stating a good open-ended question which students can investigate themselves. The next section reviews several examples of activities that are easy to adapt.

Example Activities for Inquiry

Introduction

The best way to understand how to use inquiry activities is to examine several examples of these activities for the elementary classroom. In the university/college classroom, some examples should be field-tested and students should perform the activities in the same way as their elementary school students would. In addition to discussing the actual activities, these college students should discuss how they would use the activity in their classrooms (or future classrooms), considering changes needed at different grade levels, preparation needed, sample questions, etc. It is also possible to perform one activity and discuss other activities in a collaborative discussion format.

---

3 National Science Education Standards, p. 33

4 One example of a text that includes a list of projects is *How to Teach Elementary School Science* by Peter Gega including an Appendix on " Some Major Project Developed Programs," pp.231-235.

Description of Activities

The Oil Spill

The commonly used experiment on cleaning up an oil spill is a good example of an activity in which it is easy to use the inquiry approach. Although many variations of this experiment exist, most of the prepared experiments instruct students to set up the model and test a series of specified materials. This experiment can easily be done using the inquiry approach. In this inquiry format, it is set up with an open-ended question and students are told in advance that they will be simulating an oil spill in the classroom. They are asked to think about what kinds of materials and/or methods they might use to clean up this spill, and they are instructed to bring in from home any materials or ideas that they would like to try. This immediately gets them involved in the design and they are motivated to think creatively about the activity.[5]

In class, the instructions are simple:

How can you clean up an oil spill?
What methods work well?
Is there any method that works the best?
What other observations can you make?
Keep records and be prepared to discuss results with the class.

Students work in groups and discuss their experimental design. Of course, the teacher must also bring experimental clean-up materials to class because there are always some students who forget. It also gives you, the teacher, an opportunity to ensure variety in the investigations and provide guidance for students as they investigate. Thus, some ideas are explored simply because the materials have been made available.

During the activity, the teacher serves as a facilitator and makes sure that all groups are proceeding reasonably. If a particular group needs help, more questions are asked to guide their thinking about the procedures being used and/or the concepts being formed. Occasionally, a group may even need one or two specific suggestions to get started, but this usually happens only when the students are not accustomed to inquiry investigations.

During the follow-up discussion, groups report out and reflect upon their statements, clarify what they have learned, reviewing the concepts and stating generalizations. As much as possible, teachers should encourage students to contribute the ideas for discussion. With the oil spill activity the students are able to describe several techniques and relate them to those used in actual practice.

The activity is a versatile activity allowing for easy integration with social studies and language arts, the inclusion of career possibilities, and the incorporation of technology. Technology can be incorporated in two ways. Computer programs about the environment can be employed and this is an ideal place to build concepts relating to the importance of technology in our everyday lives by discussing the actual technological methods used in cleaning up oil spills.

5 Brendzel, Sharon "Cleaning Up an Oil Spill," *Science Scope*

Solution Rates

There are many other examples of inquiry type activities. As in the previous experiment, it is as simple as stating an open-ended question and letting the students investigate. For example, in teaching about the solution rate of solids in liquids, a good question would be "What factors affect the rate at which salt dissolves in water?" Students are then given the opportunity to examine the factors that they choose after a group discussion. It is easy for the students to think of temperature, volume and stirring and begin a good investigation. Most youngsters have had experience with solutions in their everyday life and they find this an easy experiment to work with. If, however, the students do not come up with one or more factors that the teacher finds important, then, the teacher, as the facilitator, might guide the class to discover that factor as one that ought to be investigated. It should be obvious that having the teacher give the full set of directions, or list and explain the factors relevant to rates of solution, is a close-ended approach that is much less likely to be effectively absorbed by the students. The activity becomes an inquiry activity when the students begin the exploration with an open-ended question and use their own experimental design to solve the problem.

Density

The concept of density is complex, and students often have a difficult time understanding this concept. Approaching this activity in an open-ended way helps students to avoid misconceptions. While a recipe approach does lead students to the concept, it does not fully involve the students and the student is therefore more likely to forget the day's lesson. Using open-ended inquiry approaches gets the student involved so that he/she is more likely to understand and learn the concept. The potential pitfall is that the student will wind up in a blind alley. Here it is the teacher's role as facilitator to guide the students.

This type of intervention has successfully been used in teaching students about density. An appropriate question to start the subject might be "What makes an object sink or float?" As students begin the investigation, they may focus in on weight (the manifestation of mass on earth) only, i.e. ignore volume. If this happens, the teacher needs to provide two objects of equal weight but different in volume, forcing the students to consider volume. Sometimes it is necessary to go one step further and provide a heavier object that floats and a lighter object that sinks (for example, a large solid rubber ball compared to a small glass marble). Another set of objects that ought to be provided is one that includes different materials that have the same shape and volume but different mass. When the activity is presented this way, students usually develop the idea that mass, volume and the type of material are important factors. Often they have an idea that the "compactness" has something to do with it.

Although, most students (through high school) do not develop the full formula, they are, nevertheless, well on their way to understanding the concept. Often they open up other areas for future study, such as buoyancy. They also become interested in using the tools involved. If cylinders are provided to middle school children, they may look up the formula for volume of a cylinder as they investigate the properties. This particular investigation has been used successfully on a number of occasions. Thus, it should be clear that even complex questions can be presented in an open ended format. The National Science Education Standards support this approach as indicated in the following quote.

> An important stage of inquiry and of student science learning is the oral and written discourse that focuses the attention of students on how they know what they know and how their knowledge connects to larger ideas, and domains, and the world beyond the classroom. Teachers directly support and guide this discourse.[6]

The above activities exemplify the inquiry approach. They allow students to create the experimental design, and solve the problem. The teacher should guide the students as needed.

In most cases, science is also an excellent source for thematic units, allowing easily for the use of integration, technology, and knowledge of cultural contributions. A thematic unit also allows for the in-depth understanding of concepts suggested in the national standards. Thus, in the oil spill experiment, the activity can easily be correlated with social studies and language arts through articles and written work relating to various spills noted in current event articles. In the university/college classroom, students should consider other creative ways to use these lessons in their elementary classes. Consideration can be given to topics such as integration and/or other standards that the activity exemplifies.

Discussion of Correlation with Inquiry

Using the open-ended approach encourages a higher level of student involvement and opens many possibilities for future investigation. Students use their imagination and get involved in several ideas for investigation. The university/college student can discuss these possibilities, describe adaptations and consider appropriate K-8 classroom uses.

Inquiry verses Verification

For a more thorough understanding, inquiry activities need to be compared to verification activities. Inquiry can be used more often than most teachers currently realize and teaching units should include open-ended activities as much as possible. Appropriate opportunities should be used. However, sometimes, verification is appropriate or necessary, depending upon the nature of the material being taught and the learning stage of the students. The students may need to see and understand a particular science concept before they can explore and investigate. As a teacher using these activities, you will become more comfortable in choosing the appropriate technique. The chart below (Table 3.1) compares the two approaches. As indicated in the chart, verification involves the illustration of a concept.

---

6 *National Science Education Standards*, p. 34

Table 3.1 Inquiry vs. Verification

Inquiry	**Verification**
Open-ended questions	Specific questions
Student design	Recipe directions
Problem solving	Illustrates a concept

Example Activity for Verification — The Lung Model

Correlation with Inquiry/Verification

This chapter focuses on inquiry and the open-ended approach. Several activity examples have been given illustrating the inquiry approach. The lesson example below illustrates the verification approach, accompanied by good open-ended questions. Students in the university/college class should discuss various situations and describe the approach that fits best for their classrooms or potential future classrooms.

Description of Activity

A good example of a verification activity is the construction of a working model of the lungs and diaphragm. The students become actively involved using this model. The model is constructed in the following manor.

Cut a two-liter soda bottle in half, use the top half.
Take two flexible straws and attach a small balloon to the straight end of each straw using elastic or thread.
Place the straws through the neck of the soda bottle with the flexible ends near the top.
Stuff the neck of the bottle with cotton balls.
Then place a piece of plastic wrap around the bottom cut half of the soda bottle using elastic (see Figure 3.1).
Pull the plastic wrap down (representing the movement of the diaphragm downward) and the balloon lungs inflate. Push the plastic wrap up (representing the movement of the diaphragm upwards) and the balloon lungs deflate.

The model is a working model of the lungs and illustrates the movement of the diaphragm in relation to the lungs. This model is very useful even though it is not an open-ended activity, because it helps students to understand something they cannot see. Of course, it can be made more open- ended by asking good questions once the model is built.

Questions to guide students and the answers follow (answers are in italics):

Label the parts of the model.

(Students should be able to label the lungs, diaphragm, bronchi, rib cage and trachea)

How does the model differ from the actual lung/diaphragm system?

(The rib cage, represented by the plastic, does not move as air is "inhaled" and "exhaled." In the actual breathing mechanism the rib cage moves up and out with inhalation and responds in the opposite manner for exhalation. Another difference is that the trachea is actually a single tube, but it is represented in the model by two straws. This is part of the modification made when using inexpensive materials and creating two lungs. Obviously, the overall difference is that this model is made of synthetic non-living materials.)

What would happen to the system if there were damage to the diaphragm, (a puncture, for example)?

(A hole in the diaphragm would change the air pressure balance and air would not get to the lungs, or might get to the lungs in a reduced fashion.)

What other damage to parts would affect the working mechanism? How could such damage occur?

(A hole in any part of the system would cause reduced or no air-flow to the lungs. This could occur through an accident of some type or due to a genetic disorder. Another problem could occur if one of the tubes were constricted or blocked due to a growth or tumor.)

Figure 3.1 Lung Model

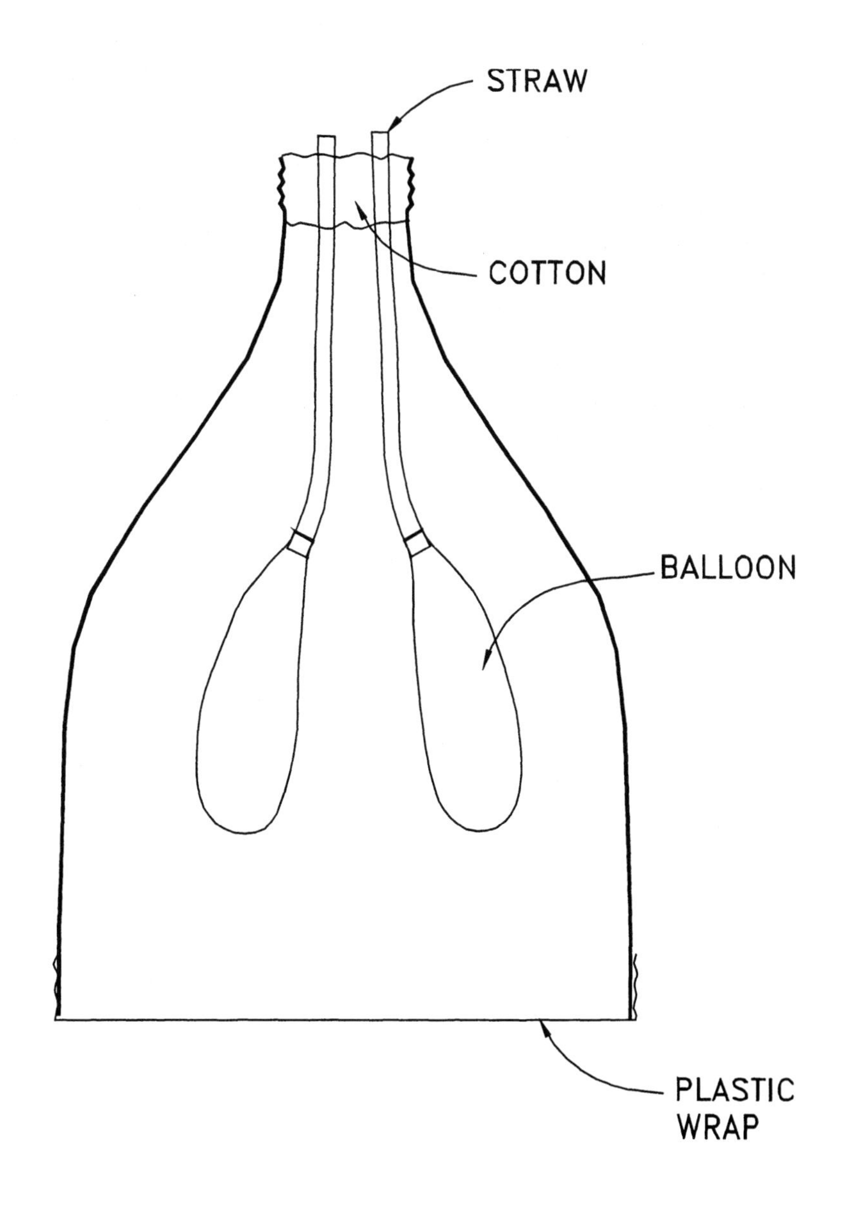

Discussion of Correlation with Inquiry/Verification

Although this activity is a verification activity, the questions that follow the activity are open-ended. Once the students have built the model, they can utilize it in thinking about the concepts. Few, if any activities need to be totally closed. Also, note that this activity can be presented in a more open, challenging way if the students are in the intermediate elementary grade level or above. The students could be presented with the materials and challenged to build a working lung/diaphragm system with the materials provided.

Concluding Remarks

To summarize, the key to inquiry is in developing good questioning techniques for student involvement. When activities are designed, this should always be kept in mind. Look over the activity to be used, and decide what portion the students can investigate on their own. This will be partially dependent on their grade level and current knowledge. Thus, when the subject matter and or grade level call for a more directed approach to begin the work, open-ended questions can be used as a follow-up

The possibilities are endless and the method is highly motivational

Selected Sample Websites

http://www.nap.edu/html/inquiry_addendum/appC.html

This is an appendix on the teaching of inquiry in *Inquiry and the National Education Standards,* Appendix C Resources for Teaching Science Through Inquiry. The connections include: Websites, Book, Journals, Resource Guides, Instructional Material and Video Collections.

http://www.education.indiana.edu

The section on Professional Teacher Resources includes the Inquiry Learning Forum with information on the following topics: What is Inquiry?; Why do Inquiry?, Connection to Standards; and How can the ILF help?.

http://www.smmm.org/sln

This site called the "Thinking Fountain" is prepared by the Science Museum of Minnesota. It is well organized and lets you approach ideas from a variety of channels (alphabetical, theme clusters etc.). Directions are given for preparing activities but thinking questions are also provided.

http://www.exploratorium.com

This website offers information about the Exploratorium museum in San Francisco. It includes lesson plans, an on-line bookstore and a program section. In the program section one of the programs is the "Institute for Inquiry" which connects you to information about books, activities and research on inquiry.

References:

American Association for the Advancement of Science. (1993). *Benchmarks for Science Literacy: Science for All Americans.* Washington, D. C.: AAAS.

American Association for the Advancement of Science. (Spring, 1995). "Common Ground: Benchmarks and National Standards." *2061 Today*. Washington, D. C.: AAAS.

Brendzel, Sharon (1998). "Cleaning Up an Oil Spill: An Inquiry Activity" *Science Scope* 22, 7.

Brendzel, Sharon (1995). "Hands-On Science for LEP Students" *NJTESOL-BE, Inc. Newsletter.*

Brendzel, Sharon and Lucy Orfan. (1995). "Integrating Mathematics and Science: A Problem Solving Approach," *Educator's Clipboard.* 7, 1.

Gega. Peter. *How to Teach Elementary School Science.* (1998). New York, Macmillan Publishing Company.

Graika, Tom. (1989). "Minds-on, Hands-on Science" *Science Scope,* (March).

National Science Education Standards. (1996). Washington, D. C.: National Academy of Science.

Chapter 4
Presentation: Units and Lesson Plans

Introduction

For each day that you teach, activities and investigations need to be presented within a concept framework that builds understanding for students. Teachers need to plan these concepts with daily activities that build towards the desired goals. Units are considered the framework for setting-up the goals, plans, activities, learning experiences and assessment of these learning experiences.

A unit is composed of a sequence of lessons used to teach a particular topic. Usually a unit lasts for one to three weeks, but the time period may vary according to the topic and the classroom schedule. The unit includes the objectives, specific lesson plans including activities, resources, and assessments. It may also include additional related materials. The components of a typical unit are described in more detail in the next section.

Although the unit is the structure or framework, the planning of the unit involves decisions and choices that comprise the heart of the teaching unit. Good choices result in good teaching. Unfortunately, poor choices result in poor teaching. You need to choose a series of lessons that provide variety, motivation, and are useful in formulating important concepts.

Frequently teachers set-up their units to correspond with appropriate sections of the text. This is acceptable if the text is well organized for the intended subject. However, teachers should always examine the text and decide how much should be used, how much should be changed, and how much should be supplemented. The newer textbooks do have many wonderful pictures, good basic activities and a variety of useful supplemental materials. It still remains the job of the teacher to properly select and use the materials.

Often the activities are recipe oriented and can easily be changed to inquiry. When possible, the activities should be re-arranged to include the inquiry investigations discussed in Chapter 3. Sometimes, the teacher is aware of a more relevant activity or one that is considered better for the current class. If that is the case, the substitution should be made. If you are a first year teacher and you are not familiar with the program or aware of other activities, use a good current text and follow it at first. However, it is essential that you analyze the success of the program and make notes so that changes can be made in the future. Experienced teachers often keep these records in their lesson plans or teacher's edition of the text.

In order to be a good teacher, it is important to be aware of what works well and what does not. You need to examine the many excellent science resources available (see Chapter 11) for good supplemental ideas and activities. As time permits, you should even look for resources and examine ideas for units that do work. Although the material you use works, it is possible that there is something even better available. Of course, you should prioritize, and work first on the areas that need the most work. The point being made here is that good teaching requires continuous growth and self-examination. New ideas blend with the old to provide the best teaching experience for your students. Stay abreast of new information and revise your units as needed.

Constructing a Unit Plan

Units represent a section of teaching and can be constructed in a number of ways. The section or division is correlated with the subject matter and the plans are related to the subject theme. Unit plans can vary but a typical unit consists of the parts listed and described below. As described above, remember to keep in mind the heart of the matter, choosing good ideas for each lesson.

Rationale

The rationale is a brief description of the unit including the reasons for teaching the unit. The reason may be as simple as the choice based upon school curricula selections, but it may be more complex in a teacher-directed integrated unit. For example, a unit on the rain forest can be created even if it is not in the textbook.

Objectives

Unit objectives can be general and need not be behavioral. The major goals or objectives of the unit should be listed. Basically, these objectives include the major concepts you expect the students to learn in the unit.

Content

Subject matter content can be included in a separate outline or the content can be included in the daily lessons. This includes the academic subject matter to be studied.

Daily Lesson Plan Outline

This consists of the list of lessons for each day. It should include the following:
Objectives
Activity
Student evaluation
All of these parts can be presented in an abbreviated manner.
The major behavioral objectives should be listed.
The activity should be summarized.

These are daily lessons, so evaluation might include observations or simple checks for understanding; tests are not usually a daily evaluation.
The combination of all daily lessons should indicate a good choice of activities, including variety and a flow appropriate for student learning. A full lesson plan is discussed in more detail later in this chapter.

Sample Lessons

When preparing these sample lessons for your own use, the short form lesson plans are satisfactory. If you need to present this unit to a supervisor, a complete lesson plan may be required. You may also choose to do a complete lesson plan for a new topic that you need to think through more carefully. The short form lesson plan should save you time in writing. It is not intended to save time in planning. The thinking process should include all of the ideas in the complete lesson plan. The short form includes the major objectives, the actual activity or body of the lesson, and the daily student assessment.

Laboratories

Laboratories are legitimate scientific experiments or investigations appropriate for the grade level. In most units one or two such activities are possible. However, this does depend upon the grade level and the unit. These laboratories should be included in the unit, either following the lesson in which they are used or in an appropriately labeled section. Cross-reference your materials satisfactorily so that you can find them when needed.

Activities

Activities include active hands-on participatory lessons. At the elementary grade level, many of these activities cannot appropriately be called laboratories. These activities may include simulations, games, paper and pencil activities etc. These activities, like the laboratories just described should be included in the unit, either following the lesson in which they are used or in an appropriately labeled section. As with the laboratories, cross-reference your materials satisfactorily so that you can find them when needed.

Resources

This section should include any additional resources beyond the activities. Examples include: computer software, audio visual aids, special kits, special references, books, etc. The resources should be those you can actually obtain and use in your school district. You might also want to make notes about additional desired resources for future use.

Assessment

Assessment actually includes two parts: *Student Assessment and Teacher Assessment.* I have chosen to use the term assessment for the student assessment and the term analysis for the teacher assessment.

Assessment (Student) – Decide how you will assess your students. Include tests, quizzes, or other written work. If the students are doing projects an/or oral reports, describe them or include the student worksheet that they would receive.

Decide how the package fits together for assessment and evaluation. More details and discussion of assessment techniques are included in Chapter 5.
Assessment

Analysis (Teacher Self Assessment)

This is just as important as Student Assessment. You need to evaluate the way things worked. What went well? What did not? What improvements are needed for the future? If you have any comments about special aspects you want to look for, feel free to make these notations.

Bibliography

This is essential for your future reference. It should include the materials you used to prepare your unit, and materials that the students will be using.

Appendix

This is a compilation of materials that are applicable for use in the unit but have not been selected. These materials have several uses in your classroom for the future. You may

- need additional activities for a slightly longer unit, or
- wish to substitute an activity at a different level, or
- need enrichment activities, or
- want an activity that is easier to modify for individuals.

Description of a Complete Lesson Plan

Introduction

Within the unit plan, each day's activity needs to be structured for student learning. This includes your objectives, the way the objectives will be achieved, and checking on student learning. All lesson plans include these components in some format. The type of format varies greatly and you can choose the one that works best for you unless the district you work in has a specific outline you must follow. You can still make that outline work for you. Many experienced teachers have the outline in their heads and use a shortened version to save time. This short form was discussed under the Unit Plan.

If you are a novice, you will need to follow a complete lesson plan until it becomes a part of you. Experienced teachers often use a complete lesson plan, for example, when planning something totally new for them. The format serves as a guideline to remind you of the important components in a good lesson. Of course, it is always important that you think about the best way to present the material for student motivation and learning. Therefore, it is important that you remember many of the ideas discussed in previous chapters and many ideas that you have previously learned about good teaching. Once you have the general idea, you organize the details according to a framework. A good sample lesson format is provided below, but remember that there are many reasonable variations of this format.

Lesson Plan

Identification

This includes date, grade level, and class period (and/or whatever is needed for your identification within the unit).

Objectives

Daily objectives need to be behavioral objectives stated in terms of what the student will be doing. You should not use the word "understand." Ask yourself what the student will be doing to indicate that he/she understands the concept. These objectives are usually action verbs such as performing an experiment, making a list, preparing a chart, drawing a diagram, explaining the concept orally or in written form, etc. For a more complete list of common verbs used in writing behavioral objectives see Figure 4.1. Be aware that this is not an exhaustive list, but these words will serve as good examples to get you thinking along the right pathways. Daily objectives should always include the cognitive objectives (content), but may also include the affective (values) or psychomotor (physical movement) objectives if appropriate.

Procedure

This is the body of the lesson, the main form of presentation or the actual activity being done. The procedure should be broken down into three parts as described below.

Introduction

The motivational set or "hook" to get the students interested in the day's work. This can be done in a variety of ways. For example, a demonstration might be used, a book or poem might be read, a good question might be posed etc. The important point is to introduce the topic in an interesting way so that students will be eager to do the day's activity. Madeline Hunter calls this the "Anticipatory Set" and that term has become popularized through her work.

Body

This includes the actual activity (used here in the broadest terms). Thus, it might include a summary of the procedure or directions for doing the activity.

Closure

This includes a wrap-up of the activity. It might be a summary, (preferably by the students or group reports from students), a few summary questions or an example of an application. If this is done orally, the ideas should come from the students but the teacher needs to add comments as needed for clarification. It is important that the students understand what they have learned through the activity and the teacher, of course, is the guide in this process.

Materials

List any special materials needed for the activity. The basics do not need to be listed (text, chalk etc.). If you are doing an investigative type of activity, this material list will be particularly important to help keep you and your students organized. Some lesson plan formats include the material list before the procedure. You should choose the location that works best for you.

Assessment

Student work needs to be assessed daily. This is not the end of unit test or assessment, but a daily check. It might include teacher observation, written or oral questions or a specific appropriate assignment such as a drawing. Keep in mind Bloom's Taxonomy, and use higher order thinking skills whenever possible. Often, applications of the day's work can be used. It might be done in class or as homework, depending on the length of the activity.

Analysis

This is the self-examination portion of the lesson, often considered the most important part of the lesson. How did the lesson go? What would you do differently next year? As indicated in the introductory portion of this chapter, notes should be made for your use the next time you present this material.

Figure 4.1 Sample Behavioral Objective Terms

Explain	Experiment
List	Conclude
State	Predict
Solve	Analyze
Graph	Diagram
Build	Draw
Measure	Label
Write	Infer
Construct	Demonstrate
Show	Estimate

Sample Lesson Plan – The Footprint Puzzle

Introduction

A sample lesson plan based upon the footprint lesson presented in Chapter One is included here. Keep in mind that this is a sample. There are, of course, other good ways to accomplish various portions of the lesson, but this model should assist you as you prepare your own lessons. It should serve as a model, not an absolute prescription.

Correlation with Lesson Planning

This chapter describes unit plans and lesson plans. A sample lesson is provided to serve as a guideline. The lesson chosen is one you should be familiar with from Chapter 1. Of course, different ideas are possible, but the example given suggests some of the good ways to use this lesson material and write it up in a correct format. In you university/college classrooms, design additional lesson plans and discuss these in your groups. Consider appropriate variations and extensions of the lesson ideas.

Description of Sample Lesson Plan for K- 8 Students

Identification

Title: Footprint Puzzle
Subject: Earth Science
Grade: Varies
Topic: Paleontology, footprints

Lesson Objectives:

The student will be able to:

1. Analyze the footprints and list specific evidence supporting statements about the footprint set.
2. List examples of the types of information that can be obtained from footprints.
3. Describe problems that might occur in interpreting records of the past.
4. Develop a scenario that adequately explains the footprints seen in the sample.
5. Evaluate the various scenarios presented and select the most logical scenarios from the supporting evidence.

Procedure

Introduction

Show students slides or pictures of footprints and ask them to explain and interpret the footprints seen. Use student responses to help develop an understanding about information provided by footprints. Ask open-ended questions, and narrow down as needed.

For example, show a particular picture and ask:

What information does this set of footprints provide for us?

If students do not respond, ask questions such as:

Can you tell the type of animal that made these footprints?

Can you tell what the animal was doing when the footprints were made?

What does the depth of the footprint indicate?

Body

Give the students the attached footprint puzzle. Break the students up into small discussion groups and assign one person as the recorder. Other specific assignments can also be made to assist the groups in functioning.

Ask the students to discuss the footprint puzzle and develop a scenario or story to explain the footprint pattern seen in the picture. Students are to be directed to include an explanation for all observed patterns. They are also to be instructed to assume that this is all of the evidence we have. In other words, this is the record as the paleontologists found it. If they wish to obtain further evidence to help support their scenarios, they may state the type of evidence they wish to gather, but they may not assume any additional evidence.

After the discussion, each group is to write a summary of the group scenario. If there is some disagreement among group members, more than one story can be described. In each case, supporting evidence is essential.

Have one student from each group summarize the scenario for his/her group. Lead a class discussion so that students understand the various possibilities and the supporting evidence.

Closure

Have students summarize the kinds of information that can be learned from footprints. Have students list the evidence needed to determine the information. Have students analyze and evaluate the various scenarios presented in terms of the evidence.

Assessment

Keep records of student participation in discussion and /or questions relating to the day's discussion or have the students write the answers to some questions. (Questions should include application, analysis, synthesis and evaluation.) For example:

1. Which scenario do you think is most supportable from the evidence seen?
2. What kinds of activities would provide footprints that could mislead the paleontologist interpreting the footprints?
3. What conditions of the ground might alter the interpretation of footprints?

Materials

Footprint pictures, slides, samples
Footprint puzzle worksheet
Overhead transparency of footprint puzzle

Analysis

Self-evaluation and analysis of lesson plan.

Next Related Lesson

Show a short film on a paleontology dig.

(This step is not included in the general description of the lesson plan above, but would be evident in a unit plan. It is always useful to think about the flow of the lessons.)

Lesson Analysis

Within the lesson plan for the footprint puzzle, several variations were mentioned. Of course, additional ideas are possible. There are many good ways to present the same material. You need to consider several ideas, and choose good examples. Select ideas that are motivating for students, keep in mind the student ability level, vary the selection of techniques, and be aware of the availability of materials. Good lesson plans result from considering all of these aspects, choosing good material, and organizing the material well for presentation.

Of course, as already mentioned, other variations for the lesson format also exist. You might choose to discuss some of the pros and cons of some alternative formats.

Discussion of Correlation with Lesson Planning

In your university/college classroom you should discuss this lesson plan and other examples of lesson plans. Consider appropriate classroom use and discuss adaptations for different grade levels. This should be done for several lessons so that you are comfortable with the idea of planning for a variety of situations.

Concluding Remarks

The essence of good lessons is in the planning and selection of appropriate activities and experiences for the content unit. The selection involves research and resources. The planning involves the use of the selected materials in an organized manner. The formats in this chapter should help you to organize your lessons so that you can follow through in an appropriate manner.

Selected Sample Websites

tikkun.ed.asu.edu/coe/links/lessons.html

This site at Arizona State University links you to a large number of sites with lesson plans including ask ERIC and a list of several other links. Each link provides a variety of materials including lesson plans with good ideas and lesson plan formats included.

www.trc.org

This site prepared by the Wayne Finger Lakes Teacher Resource Center has four divisions: Search Engine; Education; News, Weather, Sports; Reference. The Education section links you to several possibilities including lesson plans by topic.

References

Brendzel, Sharon (1999). "Prints to Ponder" *Science Scope*, 22,7.

Hunter, Madeline (1986). *Mastery Teaching*. El Segundo, CA: TIPS Publications.

Chapter 5
Assessment

Introduction

Current "best practice" includes the use of alternative assessments, using a variety of assessment techniques rather than relying on the traditional paper and pencil tests with multiple choice, matching, or fill-in-the blanks questions. The assessment should be appropriate to the activity. If you use open-ended activities, a traditional test does not fit the learning accomplished. Instead, you need to use open-ended questions or journal writing or activity-type assessments. However, traditional test type questions can be modified. For example, multiple-choice questions can be made to involve more problem-solving, rather than focusing on the lowest levels of Bloom's Taxonomy (Knowledge and Comprehension).

There are many texts and articles about alterative assessments. This review is a review of the alternative assessments most appropriate to science teaching. These alternative assessments are listed and discussed so that you can apply them in the classroom. If a more thorough review is needed, refer to the bibliography.

Alternative Assessments

Performance Assessment

In science this may include a practical type of assessment where students perform an activity similar to the one done in class. The examples you might remember from your schooling would include a chemistry practicum on identifying the unknown substance after performing many similar tests in the laboratory or a biology identification practicum after performing a dissection activity. For elementary school children, most performance based assessments would be simpler activities than those just described here. Examples that have been included in standardized tests include separating sand and soil, classifying buttons etc. Thus, for example, if students did an experiment on solution rates using salt and water, they might be asked to design an activity to test the solution rates of sugar in water. The application would be more complex for higher grade levels.

Open-ended Questions

These are questions that can be answered by short essays, appropriate for students according to the content and age level. It allows the students to summarize what they have learned in their own words. Included here are some examples related to previously used activities.

Give two examples of magnetic objects and two examples of non- magnetic objects.

Describe a good way to clean up an oil spill.

Explain why some objects sink and some objects float.

Drawings or Diagrams

These are especially valuable in science. They allow students to express their knowledge in a non-verbal manner. Science content often is easily assessed through diagrams or drawings. Before and after drawings can be very useful for the sake of comparisons. Drawings are also useful for young children who may not yet have all the verbal skills needed for appropriate communication.

Journals

Journals are daily records and can be designed by the teacher. Students can answer open-ended questions or summarize lab results or use drawings etc. For example, the journal page for the lung model could easily include some of the open-ended questions suggestion in the activity and/or a diagram of the model.

Students can also be asked to give their reactions to the activity. This helps the teacher in future activity preparations.

Laboratory Reports

Traditionally science teachers have asked students to report the results of experiments in a laboratory format. This can still be used on occasion, but the complete report usually takes a lot of time for both student and teacher. It will be more useful, for most of the time, to focus on one or two portions at a time. This gives students an opportunity to focus on a particular skill and you, as the teacher, a better chance to properly grade and guide the students. Thus, the assignment could be to simply turn in the data recorded, allowing the students to learn how to record charts. Of course, the other parts, especially the conclusion, need to be discussed orally. On another day, the conclusion only can be written for the report etc.

Portfolios

Portfolios are collections of the students work, and can be used in any subject. One of the assets of keeping a portfolio is that it encourages students to self–analyze and evaluate their own work. Usually the items to be kept in the portfolio are selected by the teacher and the student together, consisting of the student's best work. This helps the student to recognize his/her good work and to understand some of the problems in the work not selected. Also, portfolios are cumulative works in progress. Students should compare recent work to earlier work and be able to see growth. Hopefully the growth is in learning as well as in appropriate student self-evaluation.

Oral Reports

Students should have some practice with this form of communication and it is easy to incorporate it into group work, having one student report-out for the group. Special projects also lend themselves well to this type of evaluation. In addition, teachers should consider student learning styles and decide if this type of assessment is better in a particular time and/or place. Thus, for example a bilingual student may have mastered verbal skills in the new language earlier than written skills, and an oral assessment would give the student a better opportunity to express his/her learning.

One or Two Minute Summaries

A good way to assess the understanding of a particular lesson is simply to ask the student to write about the most important thing he/she learned in today's lesson. This is a valuable tool for teacher use in assessing whether or not the material needs to be covered in another lesson, and also helps you to see student differences in progress. Of course, these questions can also be part of a journal entry.

Modifying Traditional Test Types

Multiple-choice tests have typically been used to test the lower levels of knowledge on Bloom's scale. In this format, the question has one answer and usually involves recall. However, in today's standardized tests, multiple-choice questions are more complex. In your class, you can use multiple-choice questions in the same way. By using more complex questions and answer choices, the questions can test higher –level reasoning skills because they require students to think in order to solve the problem. Multiple-choice questions can also be constructed in a format where more than one correct answer is possible. In this case, students have to examine all choices, rather than looking simply for the best choice. Another variation can be to use the phrases "all of the above" or "none of the above" as a choice. Thus, we can see that there are several ways to construct multiple-choice questions to include higher level thinking skills, and students should get practice with this format.

Evaluating Alternative Assessments (Using Rubrics)

Alternative assessments should be a tool for teacher self-evaluation. Did the students understand the main concept or concepts? Was the material presented at the appropriate level? Was the activity understood? Grades do not need to be assigned. Assessment can be diagnostic.

However, teachers have also realized that alternative assessments help students with various learning styles to illustrate their learning in appropriate ways. If all tests are not written, and if different skills are utilized, students can demonstrate their knowledge and achieve at their levels. The hope and the reality is that more students are successful when you give them various types of assessments.

Of course, some of these assessment techniques are more difficult to grade than the standard multiple-choice or matching items used to test lower-level knowledge skills. In order to grade a performance assessment, for example, one needs to determine the expected responses and decide how to apportion the points accumulated towards a grade. The most commonly accepted way to do this in an organized, fair way is to develop a rubric. A rubric is simply a chart with expected achievements placed in appropriate categories, and assigned competency levels. The levels vary according to needs and preferences, usually from three to five. The lowest level is generally assigned to students who do not understand the concepts and the highest level is assigned to students who have mastered the concepts. The words used are designed to make students aware of their level, but also not to insult them. Thus words used include phrases such as "Novice" rather than "Poor."

Each level needs to be assigned to the objectives and goals expected. If a skill is being tested, the teacher needs to decide his/her goals for mastery and for each level below that.

Sometimes, this is easy because several components are included and the number of concepts understood basically indicates the competency. In other situations, preparation of the rubric is more complex, requiring an examination of all of the objectives and a decision as to the relative weight of each expected achievement. Teachers need to develop the rubrics according to their expectations, but revisions may be needed after examining students' actual responses. (As teachers, we are well aware that students often do respond differently than expected, and we need to give consideration to valid responses even if we did not think of them originally.) Developing these rubrics will take practice, but it is better to get started on this approach than to use the old fashioned method consistently. Let students be aware of your expectations, whenever appropriate, and be flexible as needed. As you prepare rubrics, keep in mind that you have done something like this when you graded an "essay" question previously. You needed to assign points to different responses and this is similar in many ways to a rubric.

Table 5.1 illustrates one example for a lesson on building a terrarium. Two objectives are used and the example rubric ranges from "needs opportunity for further development" to "outstanding."

Table 5.1 Rubric for Assessment

Topic: Building a Terrarium
Grade Level : 4

Objective 1 TSWBAT: illustrate a food chain

Level 4 Outstanding
Drawing includes: plant, animal, soil, arrows, oxygen, carbon dioxide

Level 3 Average
Drawing includes: plant, animal and arrows **or** gas component (one missing)

Level 2 Minimal Competency
Drawing includes: plant and animal but no arrows or gas components

Level 1 Needs Opportunity for further Development
One of primary components missing (plant or animal)

Objective 2 TSWBAT: develop a design for setting up a terrarium

Level 4 Outstanding
Design includes: plant, animal, soil, enclosure

Level 3 Average
Design includes: most parts (missing one component)

Level 2 Minimal competency
Design includes: half of needed components

Level 1 Needs Opportunity for Further Development
Design includes: minimal parts (basically one or parts that are wrong or confused)

Example Activity — An Inquiry Approach to the Paper Towel Activity

Correlation with Assessment

You can use any activity to practice developing rubrics, but I have chosen the "Paper Towel Activity" because it will allow you to develop content and process skill rubrics. The lesson is presented with the activity and, after completing the activity, a rubric should be developed and discussed

This particular lesson on testing the absorbency of various paper towels is a popular activity used at many grade levels because it is simple, fun, and serves as an effective teaching tool for getting students involved in the process of science. In the university/college classroom, you should perform this activity and develop a rubric for a few of the activity objectives. Discuss the advantages and problems in using a rubric for assessment.

Introduction to Activity

Although, most of the text formats provided for this experiment, direct the students and give them the procedure to use, this activity can easily be turned into an open-ended inquiry activity. The students are asked to compare paper towels by answering the question, "Which paper towel is the most absorbent?"

The students are instructed to discuss the question, and design a plan to test the paper towels. In doing this, they should make a prediction/hypothesis of expected results and they should have a plan for recording the results so that they can form conclusions. In designing the plan, students are advised to agree on an understanding of the term absorbent. For example, students may think Bounty will be the most absorbent because it is advertised as "the quicker picker upper." Students should consider factors such as the thickness and texture of the towel in making their predictions.

Description of Activity

Once students are ready to begin, the teacher provides several brands of paper towels. Be sure to include some extra absorbent samples (such as Shop Rite Super Absorbent or Viva Ultra), and some very low absorbency brands. Other materials are provided at the request of students. Some materials are set out to be used and others are taken from cabinets at the specific request of students. Typically, you should have available an assortment of beakers, cylinders, medicine droppers, rulers and balance scales.

Students choose a variety of methods to test the paper towels and are instructed to record results in mathematical quantities. When students are ready to discuss their tests and results, we should discuss the fact that we would like to compare all of the work of the various groups. Since some students have used volume, others mass, and still others have measured linearly, we set up a ranking system for comparison. Thus, students should also rank the paper towel in order with the highest number having the most absorbency and the lowest having the least absorbency. If five brands are used, the most absorbent is ranked 5.

Then students report on their results using both forms of measurement.

There are several possible methods and some of these have been written up in texts or as laboratories in the recipe format, but with this inquiry approach,

the students can devise their own plan. For example students will cut equal size pieces of each paper towel and will drop a specific number of drops of water on each sample using an eyedropper and measuring the diameter of the circle produced by the drops. There are a variety of other methods that can be used and these are described in more detail below. The following questions are good to use in guiding students:
What is absorbency?
How can you determine the absorbency of a paper towel? Discuss a few methods?
Which method did you choose to use? Why did you select the method?
What factors need to be controlled in order to compare the results?
Do you think some of the methods provide more accurate results? If so, why?
Should you always buy the most absorbent paper towel, now that you know the results? Explain.

As mentioned earlier, this experiment is very useful as an example of the process of scientific experiments. Students are doing all of the following:

- devising an appropriate plan
- controlling variables
- using quantitative measurements
- measuring accurately
- preparing a chart
- comparing and analyzing the results

Lesson Analysis

This experiment lends itself to good discussions about experimental procedures. The teacher can focus on any aspect desired. This is a good format for clarifying procedures and methods used in experimenting. Excellent discussions about controlling variables, comparing data, measurement techniques, and choice of methods can evolve from the experiment. If results are not consistent, and students wish to explain this, a few ideas can be discussed as follows. Good experimentation usually involves more than one trial so that careless errors are reduced. Also, some of the techniques are easier to measure accurately, yielding better results, and this can be considered. Students should be made aware of the possibilities for errors involved in their techniques, evaluating the techniques.

The activity can also serve as a vehicle for discussing the conclusions from observed results and the possibility of considering factors not indicated in the experiment. For example, the super absorbent towels come with fewer towels on the roll and cost more to buy per roll. Thus, the individual towel is considerably more. This cost can be calculated, providing a nice connection to mathematics. The question of cost in practical everyday use is an important factor. Thus, the paper towels that are in the medium absorbency range and are more reasonable to purchase are probably the best choice for normal household usage, where huge spills are not a daily occurrence.

Another extension can be to prepare a graph of the data collected.

One interesting technique factor that often comes up is that of controlling the size variable. Typically, in experiments, size is kept constant for the sake of comparison. In the example cited here, this was also done. However, some students have made the argument that a single sheet of paper towel can be used even if the size is not the same. The point made is that we tear the sheet off the roll and use it as a single unit. Thus, if a larger sheet helps in absorbency that is part of the test. This argument does make sense, and we should allow students to use it as long as they state their reasoning and have considered all aspects. (We want to be sure that it is not simply a careless error.)

Some of the many methods students choose to use are listed and briefly described below by categories.

Area. One of these methods was described in the student sample above where students measured the diameter of the ring created by the drops. Some students measure the actual area with this method.

Another variation of the area method includes cutting strips of paper and laying them overlapping a bucket, with the ends in water. The strips absorb the water over a set time period and the length and/or area of the strip is recorded.

Volume of absorption. With this idea in mind, some students will place the paper towel samples in a bucket and drip or pour water onto the towel until the towel can hold no more and is fully saturated. The liquid volume absorbed is recorded.

A reversal of this method has also been used. The students saturate each paper towel in a given quantity of liquid (for example, 15-20 ml). They then drip or wring out the paper towel to see how much liquid can be removed from the towel. A simple subtraction allows them to calculate the amount held by the paper towel sample.

Another variation in obtaining the volume of water held by the paper towel is to immerse the paper towel in water of known volume for a given controlled time period and compare the amount of water absorbed by each towel in the time given. Original volume minus the end volume indicates the absorbed amount.

Time. Using some of the above methods, students will measure the time it takes, rather than the volume. For example, the towel is immersed in a given quantity of liquid and the time it takes to absorb is noted. Of course, if all of the liquid is not absorbed, that would also be noted.

Another variation of this is to saturate the towel and note the time it takes for a specific amount of liquid to drip back out. (This can also be done by the volume method, measuring the amount that drips out in a given time period).

Mass. Using a specific size for the samples, immerse each sample, let it drip for a specified time period, and measure the mass of each wet towel. This can be compared to the mass of the dry towel of the same size and brand to determine how much mass was gained by the absorbency test.

General Observations. Pour a small amount (specified amount) of liquid on the tray and use each paper towel to pick up the spill. Which one appears to work the best? Does the towel fall apart? Is all of the liquid cleaned up or what portion remains? These kinds of observations might be additional for any of the techniques described above or might be used as a general, non quantitative way of doing the experiment. I would not encourage this for older students, but it would be a satisfactory way for younger students. If they think of it themselves, create the design etc., it is better than giving them the recipe. A discussion of ways to measure results could follow at their level of understanding.

Summary

When students use several different methods with the open-ended approach, the results will vary because of the form of measurement and because of the difference in approach. However, this is still a rich experience and students are interested in hearing the various methods used and evaluating the techniques. A reasonable comparison can be made by having students rank the results as described above. With this system, the variations are reduced and form clusters that can be compared. The best and worst are most easily identified with the rankings of the middle group changing places more often. This is not a problem, because reasonable conclusions can be drawn and, as a teaching tool, the process is more important than the actual results

Discussion of Correlation with Assessment

After you have done the activity, you should develop a rubric for at least two of the activity objectives. This should be done in your groups and you should compare your results with that of others in your class. Although there are no set answers for this activity, the rubric should fit in with the lesson objectives and should have categories distinguishing the various levels you choose.

Concluding Remarks

Assessments are an important part of good teaching and need to flow from the teaching. Choosing them appropriately is an important part of instruction. Hopefully, this chapter has given you a starting point to begin this process.

Selected Sample Websites

www.miamisci.org:80/ph/lpexamine1.html

This site gives examples of alternative assessments using ph indicators as the topic example. Examples are given for performance based, authentic or project, portfolio, and journal. The examples are illustrated.

ericae.net

ERIC Clearinghouse on Assessment and Evaluation provides a large resource network on assessment including articles, tests, resources etc'

References

Brendzel, Sharon (2002). "Science on a Roll, Part I: Absorbing Inquiry" *Science Scope.*

Brendzel, Sharon (1994). "Active Assessment for Active Science: A Guide for Elementary School Teachers. George E. Hein & Sabra Price. Portsmouth, New Hampshire, Heinemann. *Educational Studies, 26,3.*

Hein, George. E and Sabra Price (1994). *Active Assessment for Active Science: A Guide for Elementary School Teachers.* Portsmouth, NH, 1994: Heinemann.

Luft, Julie. (1997). Design Your Own Rubric (1997*) Science Scope*, Feb. 1997.

National Science Teachers Association (1994). Assessment Issue *Science and Children*, 32,2.

Chapter 6
Integration

Introduction

Science can easily serve as the theme for integrated units because all topics typically studied in a science class are naturally integrated with one or more other subject areas in the real world. The separation of subject areas has been set artificially for "convenience" in teaching curricula in schools and, this artificial separation has often resulted in student misconceptions. Specifically students taught this way often see the world as compartmentalized. Thus, educators have re-introduced the idea of integration in order to simulate the real world and help students to understand the natural connections.

Although science and mathematics are naturally linked in many real life experiences and many classroom activities, teachers often avoid the connection or fail to see the possibilities. In addition, many individual subject exercises are fabricated for students, missing the real-world connection. Today connections are being stressed and this integrated emphasis is included in the national standards for all subject areas. This chapter discusses integrating science with all subjects. Since science and mathematics are most naturally integrated in many activities, we begin with mathematics.

Integrating Science and Mathematics

The connections are often obvious and easy to do. For example, several content areas are studied in both science and mathematics including measurement, graphing, formulas etc. In addition, many experiments readily include mathematics. Examples include plant growth (measurement and graphing), density (measurement and formulas), mapping (measurement and scaling) etc. When students use the recommended problem solving open-ended approach in science activities, frequently the mathematics evolves naturally. For example in the experiment discussed earlier on solution rates (What factors affect the rate at which salt dissolves in water?), students naturally use mathematics to measure the temperature and time the trials. Recording data and keeping tables are also a natural part of these techniques, and often come from the students themselves in their group discussions about the problem.

A slightly more complex investigation involves an adaptation of the popular "Geological Time Line" investigation. Typically, texts include all of the directions for preparing a time line. Instead, ask students questions like these: "Where is human kind's place in the history of the earth?" "How can we set up a time line?" Again, give students general directions and let them devise a plan. This involves developing a scale based upon the space available and the geological data. Obviously, mathematics is an integral part of the activity.

The advantage of an approach in which students truly explore is that the students become the investigators. They feel involved and learning takes place. In the process, integration of mathematics and science occurs naturally.

Integration with Other Subject Areas

Science topics are also easily integrated with all other subject areas including language arts, social studies, art, music, physical education and world languages. Sometimes one thematic unit can be used as the central theme for all subject areas throughout a specific unit of teaching. The natural sciences easily lend themselves to integrated themes. Some examples include environmental science, natural habitats, studies of plants and animals, geological features etc. For example, a unit on the rain forest easily integrates all subjects.

As already stated, integration is more natural to real-world situations and is usually more interesting for students. When whole unit themes are not used, appropriate integration should still be used in some lessons. A discussion of elementary subject areas is included below. For each subject area, there is a discussion of general ideas for easy useful integration and ideas for some activity suggestions.

Language Arts

All areas of language arts can be integrated including the use of appropriate literature, trade books, poetry, compositions, different writing styles, oral presentations, and even word usage or spelling. Some of these areas are basically self-explanatory. For example, a unit on oceanography can easily include stories, poems, trade books, written assignments etc. Newer textbooks often correlate their units with language arts ideas and/or specific grade-level text readings. Natural themes are particularly easy to integrate with appropriate fiction, poetry, etc. Any library will offer a variety of stories featuring weather and climate, for example. However, there are some particularly good stories and you will want to begin to keep an appropriate grade-level correlation list. You may also find good readings, stories etc. related to some of the other science areas; keep a list of these harder to find stories (building bridges, for example).

Keep in mind that oral reading allows you to use books at both lower and higher grade levels with good references. Most children have read Dr. Seuss books, and you can read excerpts from the appropriate books to children at higher-grade levels, asking them to recall their childhood readings and explain the relevance to the situation. Some series are specifically geared toward this integration such as the *Magic School Bus* series. With an open mind and a little imagination, you will find more than you need, and will even need to select and choose the best ideas.

Social Studies

Social studies is almost as easy as language arts to integrate with science. Most science topics have a related cultural, economic, political or geographical aspect. The specific selection, of course, depends upon the content area of both subjects. Thus, for example, if you are studying earthquakes, you will definitely discuss the geographical location of most earthquakes (the Pacific "ring of fire"), the human response to earthquakes, and their formation with relation to tectonic plates and other relevant scientific data. If you happen to be studying the Mediterranean area, the second largest earthquake activity region in the world, a more thorough integration between science and social studies would naturally be included. General activities that easily blend science and social studies include role playing, debates, demographic studies, technology and culture, production and industrialization, weather, maps (especially topographic maps) etc.

Art

Again, this is a subject with easy and useful integration with science. Building models is an excellent way for students to gain a better understanding in most science subject areas such as animals, plants, human biology, simple machines, geology, physical science, and chemical science etc. Often these models can be made from simple materials such as pipe cleaners, popsicle sticks, felt pieces, and styrofoam. In addition, diagrams, murals, dioramas, and appropriate drawings may be used. Young students especially enjoy the use of arts and crafts to help make ideas concrete. However, projects can be developed for all grade levels. As the teacher, you need to consider the learning value of the project in relation to the time spent. If children are in the early elementary grades and the project is truly a combined art and science lesson, more time can be used. For example, students can create felt animals and learn some basic body parts.

For older students, one needs to be sure that the time spent fits with the concept being learned. There are still many worthwhile activities, such as building a clay model of the earth's structure. It should be noted that even high school teachers sometimes assign projects using models. A classic example is the assignment to build a toothpick bridge as part of a physics class.

Physical Education

Active games using scientific themes can often be found or adapted. Many games are suggested in activity books utilizing environmental or other natural themes. Adaptations can be simple such as turning a game of tag into predator-prey relationships. The usefulness of the game in teaching the concept is evaluated and the time being spent should be part of that evaluation. Longer, more complex games should have included more detailed concepts. Of course, students do enjoy the physical activity and an active game does provide for a variation from the normal teaching mode.

Music

Everyone knows that young children learn well using songs, and many songs with science themes are available. Children and teachers often enjoy making up their own songs. Naturally, a unit on the content topic of sound, should include music for a better understanding of the concepts. This is a form of natural integration which should not be overlooked. Older children may enjoy the use of music as appropriate when a particular song happens to be popular and the words relate to a thematic topic. Older songs can occasionally be used for topics where folk songs exist. Again, some variety is offered by using different media and having students discuss the concept with the song as a starting point.

World Languages

As children learn a new language, physical models and pictures are very useful. The techniques used for teaching science are thus appropriate. Participatory activities where words are associated with an action also enable children to learn the new word more easily.

Example Activity — Mariculture (An Integrated Theme Unit)

Correlation with Integration

This chapter on integration gives suggestions for integrating science with all other subject areas. Thus, the description of the activity section follows the same model. It is not a single sample lesson plan, but a description of the topic (mariculture) and the natural connections that this topic provides with all subject areas.

Introduction to Activity

Mariculture is farming the sea. Fish, shellfish and algae are raised and harvested from the edges of the ocean. The same techniques are also applied to freshwater areas (aquaculture is the term used for cultivation in all water areas).

Description of Activity

This unit topic is usually interesting to students. Frequently, current events report on some aspect of harvesting food from the sea. The unit can easily be part of an earth science course (especially in the section on oceanography) or a biology course. The topic is a natural for integration between science and social studies and all other areas can easily be brought into the unit.

The connection between science and social studies can easily be used in a research or debate format. Fishing rights and fishing territorial rights are a "hot topic" of debate between different countries. Students can debate the limitation laws and/ or can take on the role of a particular nation to discuss the consequences from that nation's point of view.

Language art activities are easily incorporated into a unit about oceans. Students can write an essay about their personal experiences, can research a specific topic, or use an outline format. Stories are abundant and lessons using various writing skills are easily incorporated. Music and crafts are also easy to incorporate. Songs are easy to find and creating animal models or building ocean dioramas is easy and fun to do.

Mathematics can be incorporated through statistics, graphs and charts. It is also used in some of the appropriate hands-on activities. One good example is the use of the well-known goldfish respiration activity. To do the activity:

- Measure the rate of respiration (by counting the gill movements or mouth movements of the goldfish over a specific time period).
- Then add ice cubes to the water in the bowl, record the new temperature, and count the movements.
- Repeat this procedure several times.
- Chart your results.
- What conclusions can you make about the effect of temperature on goldfish respiration?

You should conclude that the respiration rate of a goldfish slows down in cold water. This laboratory illustrates the ecological balance in a community based upon temperature. Students are guided to relate this activity to mariculture by considering the implications of the limiting factor of temperature in relation to fish culture. This discussion also provides the opportunity to discuss other limiting factors in terms of mariculture.

Discussion of Correlation with Integration

Mariculture is one of many good theme units for integration. Science topics provide the basis for many excellent theme units. Even if commercial materials are not available, you can create your own materials including models, slides, demonstrations etc.

Concluding Remarks

Integration is an easy natural way to teach. The flexibility of elementary classroom structure allows for this without difficult structural impediments. Ideas are plentiful, and children enjoy the integration. Start by looking for natural connections, and increase your creativity by keeping open mind. Over time, you will build up a resource list of additional new good connection ideas for your class.

Selected Sample Websites

educate.si.edu/

Smithsonian Center for Education in Museum Studies has a website with many teacher resources including: Media Catalog, Lesson Plans, Resource Bank with connecting websites, Teaching Toolkit , Professional Development and others. Many of the lesson plans are interdisciplinary and include a series of lessons. The main topics are science, language arts, and social studies and many interconnections in these areas.

whyfiles.news.wisc.edu/index.html
This is a weekly online magazine sponsored by NSF with current science topics in the news. You can search for a particular topic and read the related articles. (Students may also be able to read these articles, depending upon the grade level.) Frequently the articles give you ideas for integration because of the way they are written. For example, doing a search on the rainforest yields many articles. The first article contains information about ceremonial plants and customs and the problems of rainforest destruction suggesting integration between science and social studies.

//curry.edschool.Virginia.edu/~tgt3e/skies/
This site provides a specific example of integration for upper grades with the topics of astronomy and literature. Reviewing the ideas might spark some ideas of your own.

References

Brendzel, Sharon, Lucy Orfan and Robert Schuhmacher (2000). "Float it Down the River" *Science Scope*, 24,2.

Brendzel, Sharon (1999). "Prints to Ponder" *Science Scope*, 22,7.

Brendzel, Sharon (1999). "Questioning Science" *Curriculum Administrator*, 35,3.

Brendzel, Sharon (1998). "Cleaning Up an Oil Spill: An Inquiry Activity" *Science Scope* 22, 7.

Brendzel, Sharon. (Fall 1997). "Implementing the New Jersey Core Curriculum Standards for Science." *School Connections*, 9, 2. Kean College, Union, NJ.

Brendzel, Sharon. (1995). "Model Thermometers" (Helpful Hints Section) *Science and Children* 32, 1.

Brendzel, Sharon. (1995). "Hands-On Science for LEP Students" *NJTESOL-BE, Inc. Newsletter.*

Brendzel, Sharon and Orfan, Lucy. (1995). "Mathematics and Science: A Problem Solving Approach" *Educator's Clipboard.*

Brendzel, Sharon. (1994). "Keep Your Classroom Current: A Mariculture Model" *Science Scope* 18,1.

Brendzel, Sharon. (1994). "School Yard Erosion and Terrain Studies" *Science Scope* 17,7.

Brendzel, Sharon (1994)."Active Assessment for Active Science: A Guide for Elementary School Teachers. George E. Hein & Sabra Price. Portsmouth, New Hampshire, Heinemann. *Educational Studies, 26,3.*

Chapter 7
Creative Teaching Strategies

Introduction

As mentioned in several of the other chapters, variety helps to keep student interest and assists you in selecting the best approach for a given situation. Some of these approaches have been mentioned in previous chapters, and they will be discussed again as needed. Other approaches have not yet been discussed properly and this chapter will help review these techniques. Technology is an especially important, relatively new area, and it is important enough to warrant a separate chapter. Therefore, it will be more thoroughly examined in Chapter 8.

The list of additional teaching techniques is extensive. Think of all the good teaching techniques you use in other subject areas and most of these can be applied to science content as well. There are also some techniques that originate with science teaching and these techniques are particularly useful. Figure 7.1 includes a list of some of the strategies most often used in teaching science. This chapter includes a discussion of each strategy in terms of best practice. Strategies thoroughly covered in other chapters are cross-referenced in this chapter.

Figure 7.1 Techniques for Teaching Science

Inquiry Activities
Verification Activities
Demonstrations
Models
Crafts
Games
Current Events
Role Playing
Debates
Cooperative Learning Groups
Computers
Audiovisual Techniques
- Laser Disc
- Films
- Slides
- Tapes

Research
Integration with all subjects
- Math
- Art
- Language Art
- Music
- Social Studies
- Physical Education

Discussion of Techniques for Teaching Science

Inquiry Activities

These important activities are discussed in Chapters 1 and 3.

Verification Activities

This type of activity is discussed in Chapter 3.

Demonstrations

A laboratory or a model can be used to demonstrate a scientific concept and this is one of the most useful science teaching techniques. However, careful consideration should always be given as to when to use the demonstration and when to use the actual hands-on participatory activity or laboratory investigation.

Several factors will usually help you decide: safety, cost of materials, visibility of the demonstration, surprise factors, and time involved. Safety is always a paramount concern and activities involving heat or other potentially dangerous materials should be done as demonstrations for young children. For older children, the laboratory set-up will determine whether safety needs are met.

The cost of the material may make a demonstration the better choice, considering all factors. This is especially true when the demonstration uses an expensive piece of equipment such as a seismograph or oscilloscope. By observing a demonstration, students can get an idea how the operation takes place, even though they are not actually participating in the demonstration.

However, if the material or equipment is too small to be visible, a hands-on activity is better. Demonstrations work best with large, visible objects.

The surprise factor is part of discrepant event presentations. When students observe something unusual, their reasoning might be influenced by being too close to the materials. A good example of this is the sinking ice cube demonstration. Ice floats in water and sinks in alcohol. Showing students two tumblers with clear liquids and ice cubes that behave differently allows them to infer what happened. If the alcohol is under their nose, the smell will give it away.

Time, of course, is always an element. In spite of the fact that it takes longer to do hands-on activities, these activities should be used for all of the reasons discussed throughout this text. However, you, as the teacher, always need to choose which concepts to present as hands-on and which to present as a demonstration. Thus, time enters into the picture as part of the decision. A particular activity may be understood almost as well through a demonstration, leaving you time to do a more important participatory activity.

An example of this would be in the classic ball and ring experiment/demonstration. This is used to demonstrate the expansion of metals with heat. The ball easily passes through the ring at room temperature. Heating the ball results in expansion and the ball cannot slip through the ring. This is relatively easy for students to see, and the concept can be understood from the demonstration.

Of course, all factors must be evaluated in light of the age of the children and the classroom conditions. If the children are very young, some activities easily performed by their older counterparts will be chosen as demonstrations. If the classroom is not set-up safely, you will need to do more demonstrations. The balance of these choices is in your hands. Review all factors carefully.

Models

Models are useful for several reasons. Physical models allow us to visualize something difficult to see, especially in science where the physical construct may be two small (e.g. cells), too large (e.g. solar system models) or too far away (e.g. volcanoes), or simply too difficult to see (e.g. internal body parts or the inside of the earth). A model of the lungs was used as an example of verification in Chapter 3. In this model, the student preparation time is minimal and the demonstration model purchased from a supply house is expensive. In addition, the visibility is much better in the student-made model.

Models may also be formulas that help us to understand a physical concept such as gravity, forces, chemical reactions etc. These can be developed by the students or applied to the lesson. Of course, you will need to consider the difficulty of the concept in relation to student grade level and prior knowledge.

Crafts/Art

This topic was briefly mentioned in connection with integrating science and art. Students can build a model or create drawings or murals. Craft materials can be used too create a model (as described in the above section). The learning is an outgrowth of the craft. Thus, a model of an animal is used to teach body parts. A working model of the lungs is described in Chapter 3 and referred to above.

Drawings can include scale models so that students are aware of the proper proportions of the system being studied. Review the material to be studied and choose an appropriate creative activity. Good examples include the Geologic Time Scale, the Solar System etc. Artistic drawings, diagrams, and murals can also provide visualization of scientific concepts. Some of these art projects also provide a motivational experience for students. For example, students can create a poster or brochure related to a unit of study.

One particularly good artistic activity combines art with the environment. Students draw a mural of the rainforest on the chalkboard (or on a large piece of butcher paper). The mural is left up for several days as the introductory material is presented. When the portion of the unit dealing with destruction and clearing of rainforests is being discussed, the rate of destruction can be correlated with the removal of an equivalent portion of the mural. Thus, a statement is made that 20% of the world's tropical rain forests were destroyed between 1960 and 1990, and 20% of the mural is cut away or erased to represent this loss. Students feel ownership because of their artistic efforts and have a better comprehension of the meaning of destruction.

Games

Games can be motivational for students even when you are using them to review lower level content knowledge. The questions become part of the game setting and students are more interested than they would be in a traditional review. Many game variations are used by teachers including copying the current game show, copying a simulated ball game, and/ or constructing a game review.

In addition to this, standard games can be adapted to science, either for review or for learning the concepts. For example, a science form of bingo is a popular adaptation and works particularly well with topics such as vitamins, minerals, or chemical elements. A game like categories can, of course, be easily adapted to any subject area. Active games can also be adapted, i.e. a game of tag can become a predator/prey relationship game. Think of some traditional games and consider how you can adapt them to science.

In addition to all of the above ideas, many organizations and workshop presentation have prepared specific games to teach specific concepts. One example, used in several different but similar formats is an energy transfer/food chain game. In this game the students form two relay teams. The first student is the sun and he/she starts with a large handful of chips (or popcorn or beans). The student has to perform some simple task (such as running to a specific point and back) before handing over his chips to the producer who, in turn, performs the same task and passes the chips to the first order consumer etc. etc. The winning team is determined by a combination of time and/or chips (this varies with the different examples). You can set up your own rules. For example, the team that finishes first gets a set amount of points, and points can also be allotted to the chips. Thus, if Team A finishes first but has 0 chips left, and Team B finishes a few minutes later with 75% of the chips, Team B can be the winner. Of course, this needs to be set up in advance. Whatever format you use, be sure to review the ideas of energy transfer lost at each step. In other words, emphasize the concept.

There are many excellent examples of these types of games. They are especially abundant in the environmental science area. As you use a variety of resources and come across these good activities, collect them and add them to your resource file for later use as a teaching strategy that adds variety and spice to your classroom.

Current Events

Science is in the news daily, and one of the many reasons that science is so important is that all of our future citizens will be faced with many different and some difficult science related decisions during their lives. Often students enjoy discussing the current topics. This is especially true when the topic is truly relevant. Even college students appreciate a relevant topic. The year of the Chernobyl disaster I was teaching an introductory Geology course. The topic of nuclear energy and radiation was in the latter part of the text, and the disaster occurred near the middle of the semester. When I re-arranged the order of the syllabus to discuss the current topic as it unfolded, the students were excited (and unfortunately amazed, claiming that professors rarely paid attention to the outside world). In my view, world events create "teachable moments." Use them appropriately.

Besides keeping your eyes and ears open, there are several magazines and weekly papers that provide interesting reading (see Appendix E). Students can also be assigned to look for current events in the newspaper. The only problem with this activity is deciding how much time to devote to it. Although it is often interesting, you, as the teacher do have to decide how to achieve the balance between time well spent and time wasted. This problem is addressed later in the chapter.

Role Playing

Role-playing is often used in social studies curricula. In some science areas, it fits very well. For example, a number of environmental issues can be examined through scenarios in which students play the roles of different members in a specific situation. When appropriate, role-playing can provide a different teaching strategy. This technique is particularly useful in examining issues where the public is involved in decision making. For example, choices need to be made about trash disposal, recycling wastes, purification of water, water usage etc.

Debates

Many areas of science present themselves as issues for public debate and students can examine issues of this type in a debate format instead of the traditional research type format. Again environmental issues lend themselves well to debates, but there are other politically "hot" issues as well in the areas of biotechnology, energy, medicine etc. Choose the topics with the maturity of your students in mind. Some of these topics are upsetting enough not to be presented to very young children in a debate format. However, it is certainly useful for youngsters to explore the different sides of an issue. Choose topics that appear appropriate and the children will present the various points of view for their own analysis.

Cooperative Learning Groups

Children learn well in small groups. We have learned this informally from observation and formally from studying some of the experts in the field. Science activities have utilized this approach for years, long before the formal application became popular.

Even for answering questions, children often do better working in groups, sharing ideas, discussing, arguing, teaching each other etc. You can apply some of the formal ideas of cooperative learning and assign points to help the groups function better, but, even, if you do not use the full techniques, you should allow the students to work in groups on projects, activities, and questions. You can still hold the individuals accountable in the way you deem appropriate, but the learning will be better because of the participation of the students and their time-on-task.

Computers

This has become an essential teaching tool, and it will be discussed more thoroughly in the next chapter on technology (Chapter 8). Suffice it to say here, that the best use is for interactive programming, a format that students cannot easily obtain from the blackboard or research or other more traditional types of teaching.

Audio Visual Equipment

This category includes laser discs, films, slides, tapes etc. The best use for any teaching device should be its use in providing an option not available through other sources and this is especially pertinent to the discussion on audio-visual devices. These audio visual devices often supply us with pictures or sounds that are not easily available in other ways. For this use, they can be wonderful tools. The student studying a volcano who has never seen one should see a film version of volcanic activity. It is much more appropriate than a still photo. However, our youngsters today have many chances to see and use videos, so it is not an automatic asset to turn one on in the classroom. In fact, a long film may be viewed by students as a chance to take a nap. Select carefully, preview the tape, and prepare appropriate questions or assignments to keep students on task.

An important point to remember is that you do not have to show the entire film, slide set etc. Select the relevant portions and show just what you need. This includes the old-fashioned devices that may still be in your schools. Filmstrips do not have to be shown with the pre-set accompanying audio versions. You can select the pictures you want and just use them. However, be open-minded. Some teachers have reported that their students enjoyed one or two of these old-fashioned filmstrips because the use of filmstrips is so old that the concept is new to them. Of course, if you choose to try this approach, choose a filmstrip, or other audio visual, that you actually think is worth the trouble, and monitor reactions carefully.

Another use for some of the older audio-visual devices relates to their availability. If your school district is on a shoestring budget, look around for some of the older materials and see how they can be put to use. A number of teachers have been able to find and use old film loops, opaque projector etc. Each of these items has some use, especially if your school district doesn't have the newest most modern versions. Examine the equipment and decide if it will help you project an image better (an opaque projector might just do that) or provide a good activity for your learning center (a film loop would be excellent for this purpose). This comment about making use of resources will appear again in the chapter on resources (Chapter 11) because the same concept applies to old kits, books etc.

Of course, if your school system is fortunate enough to have the modern devices, use them and see if you need any supplemental material. If you have the newest laser disc version of elementary science, you probably have all the images you need and simply need to become familiar with the equipment.

Research

Research is important in most subjects. Children should have experience with selecting a topic and learning how to find out about it. Sometimes, the topic can be assigned or partly assigned. For example, each group studies one biome and reports to the class. The research should include the most up-to-date computer technology and the use of the library with books and magazines etc. More detail on the computer-related research will be included in the next chapter on technology (Chapter 8).

Integration

Successful integration can be achieved with all subjects (Math, Art, Language Art, Music, Social Studies and Physical Education). This is discussed in Chapter 6.

The Time Balancing Act

With all strategies and activities, the time spent must be weighed against the expected learning outcomes. This has led some teachers in the past to assume that lecture is the best method of efficiently dispensing information. Of course, we have now learned that the retention rate for lectures is low, and, conversely, the retention rate is high for active participation. We have also learned that it is better to have students experience some depth rather than a shallow coverage of many, many topics (one of the chief criticisms of our current teaching curricula). Clearly, it is better to have motivated students than bored students.

Nevertheless, there is a balance to be achieved. Activities can become simply play, current events discussed for a full period every week can be viewed by students as a way to get "off the topic," craft projects can use an inordinate amount of time on the craft or art portion, etc. Your decisions need to be based on all of the factors involved. Each unit should have some in-depth activities, and a variety of activities to assist student learning. In analyzing your lessons, ask yourself if the students were able to understand the concepts. Is more time needed, less? How can this be accomplished for the next time the lesson is taught? Was most of the time spent on task and spent towards achieving the main objectives? Questions like these will help you to make good decisions for the future.

Example Activity — Create Your Own Mock Thermometers for Student

Correlation with Creative Teaching Strategies

Throughout this chapter several examples of activities have been suggested as each strategy was discussed. The example given in this section is an example of a model thermometer for use with young school age children

Introduction to Activity

Today, teachers are urged to use "manipulatives" and hands-on experiences for students because involvement in such activities improves learning and retention.

Learning to use a thermometer is a basic skill for science and mathematics, and this skill is also used in our everyday lives. Therefore, it is useful to create a mock-thermometer which can be used as a tool for ease in learning.

Description of Activity

A simple mock thermometer can be made for easy classroom practice. Only the following inexpensive materials are needed:

red ribbon, approximately 1-2 inches wide (2.5 to 5 cm)
white ribbon, same width
needle and thread

oaktag or cardboard
ruler
marker

Directions are given following Figure 7.2

Figure 7.2 Model Thermometer

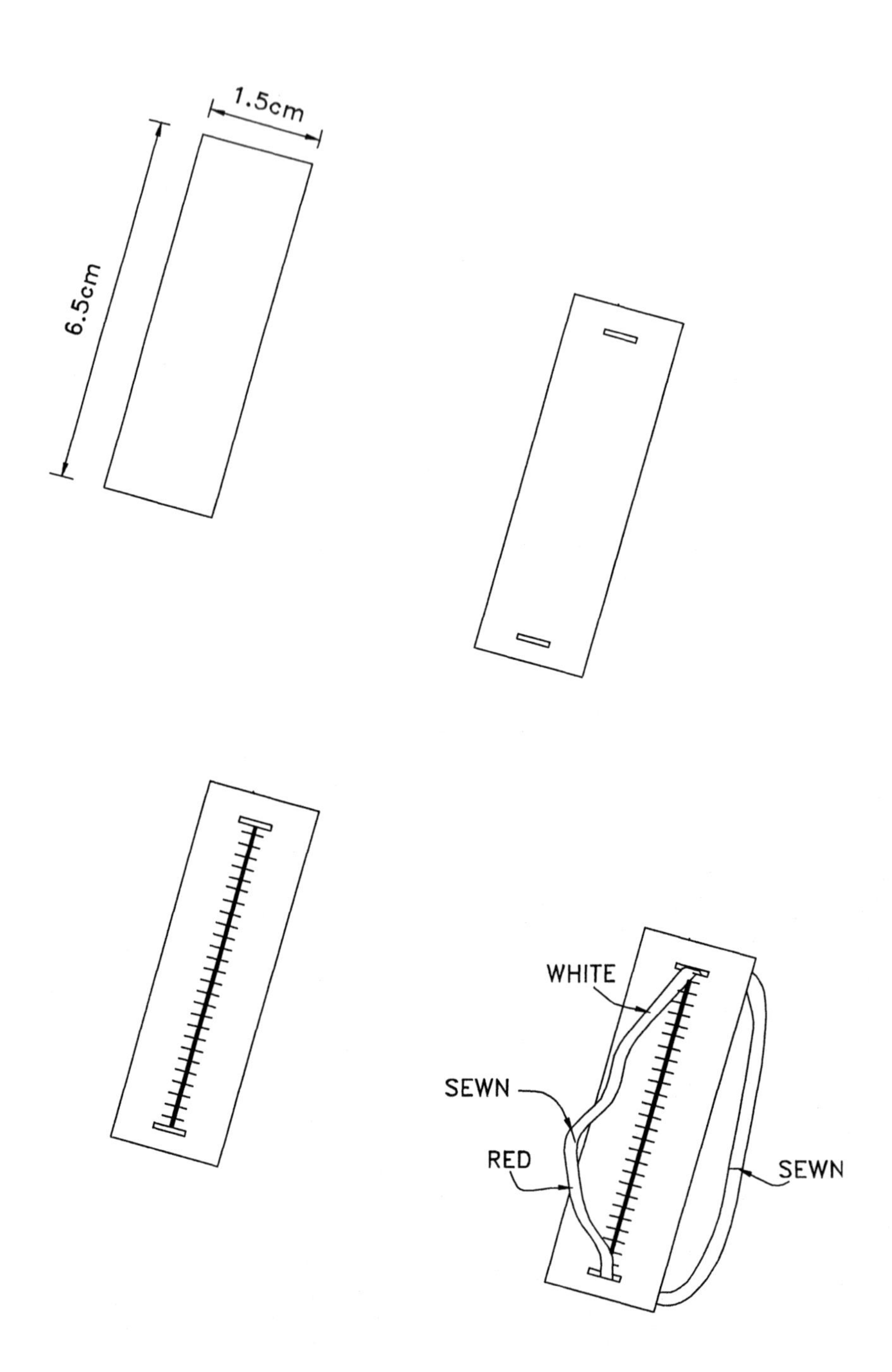

(Directions on following page)

Directions for Figure 7.2:

1. Cut the cardboard into rectangular sections of reasonable length. A convenient scale to use is 1 cm for every 2 degrees C, which translates into 50 cm for a 100 degree scale. With extra room on top and bottom to allow for the ribbons (described below), it turns out that a cardboard measuring 65 cm by 15 cm is quite adequate.

2. Cut slots for the ribbon near the top and bottom (approximately 4 cm from each end). Each slot should be as wide as the ribbon, fairly thin, and centered about the cardboard.

3. Draw a vertical line down the center of the cardboard, starting about 6 cm from one end, and then draw short horizontal lines (about 4 cm long) every 1 cm. Thereafter, lengthen every fifth line to 6 cm long, and place numbers alongside. Label the longer lines in increments of 10; that is, 0, 10, 20 etc. If you have a red marker, you may wish to draw a red circle at the bottom to imitate the "bulb".

4. Cut two lengths of ribbon, one white, one red. Each ribbon piece needs to be a little more than the distance between the slots (perhaps 4 cm more).

5. Sew the red ribbon end to the white ribbon end, thread the loose end of the white ribbon through the top slot, thread the loose end of the red ribbon through the bottom slot, and sew the two loose ribbon ends to each other in the back of the cardboard. The mock-thermometer is now complete. The "thermometer reading" is the place along the scale marked on the cardboard where the red ribbon is joined to the white ribbon. The reading can be changed by pulling on the ribbon (up or down) in the back of the cardboard.

Lesson Analysis

Once the model is finished, students can use the model to practice reading temperatures. They can be given sample problems and use this mock thermometer to practice with. For example, the teacher can set the temperature at 30 degrees and ask students to read the temperature. Conversely, the teacher can ask the students to set the thermometer at a given temperature, say 42 degrees. Exercises that involve incremental temperatures are also possible (e.g., "this morning the temperature was 24 degrees, but by mid-day the temperature climbed by 7 degrees").

The model is large enough to use for classroom demonstration and small enough to use at a table. One model can be used, or several can be used for smaller groups of students. The models can be used with or without actual thermometers to practice problems.

Discussion of Correlation with Strategies

The ideas for creative science teaching are numerous. This chapter has described many of these approaches. However, you, as the teacher, should be able to suggest many additional examples including modifications of strategies suggested here. The pros and cons should be considered as well as time, safety and cost.

In the university/college classroom, as a group, choose a strategy to discuss in more detail. Review the classroom use, and describe some additional examples for discussion with your classmates. Alternatively, a specific topic could be chosen and examples of several good strategies could be discussed.

Concluding Remarks

No matter how long the list of teaching strategies is, and no matter how thorough the discussion, other ideas will still be available. A good teacher needs to be creative and look for new ideas and new ways to approach the material. The discussions presented in all of these chapters give you some excellent tools to start the teaching project. The doors have been opened. Of course, you need to continue the work.

Selected Sample Websites

www.aps.edu

This site is sponsored by the Albuquerque Public School System, and there are many sections and links to help all teachers. Once you select Teacher Resources you are linked to several sites. For example the section on Substitute Teacher Resources includes tips and strategies for substitute teachers to take with them to class. These strategies are also good ideas for regular teachers who may want a reminder of the variety of ideas one can use and adapt in the classroom. The large section links under Teacher Resources include: General Reference, Technology Integration, APS Training Material, Kid Stuff, Curriculum Related Resources, and Substitute Teacher Resources. Each section provides additional links and many of these provide information for topics discussed in other chapters of this text. For example, the Technology Integration section provides many ideas and resources to integrate technology into the classroom.

Use a search engine. Example

www.yahoo.com/Science/Education/K_12/
This enables you to make connections with numerous other sites and gives you the opportunity to gather ideas from a number of sources.

www.hmco.com/hmco/school/search/activity2.html
This site is sponsored by Houghton Mifflin and allows you to enter through Education Place. You are then offered subject area selection. Once you select Science you have the following choices: Best of the Net, Data Place, Link Library, Project Center, Graphic Organizers, Invention Convention, Science Library Adventures and Professional Development. Each section offers ideas and examples for teaching.

References

Brendzel, Sharon (2004). "Games That Teach" *Science Scope* 27, 8:32-33.

Brendzel, Sharon. (1994). "Keep Your Classroom Current: A Mariculture Model" *Science Scope* 18,1: 32-35.

Brendzel, Sharon. (1995). "Model Thermometers" (Helpful Hints Section) *Science and Children* 32, 1: 31.

Brendzel, Sharon. (Fall 1992). "Science Education at a Crossroads," *Educator's Clipboard* 4, 2.

Wright, Richard T. and Nebel, Bernard J. (2002). *Environmental Science: Toward A Sustainable Future.* Upper Saddle River, New Jersey: Pearson Education.

Chapter 8
Educational Technology and Technology Education

Introduction

It is essential to stress that both educational technology and technology education are important parts of teaching science. All too often teachers think only about the educational technology and not the technology design elements. For clarification let's review the definitions of the two components. *Educational technology* is basically technology that we use to assist us in teaching. Today, the most important element is the computer technology including all of the facets it provides. *Technology education* is education in the way technology works, how science is applied to design and construct the components that assist us in our everyday lives.

Educational Technology

In the previous chapter there was a brief description of some of the audio-visual technology forms that assist us in teaching. In this chapter we will concentrate on the computer and discuss some of the many applications to good science teaching. For each application, we will review the usage and some of the problems because, of course, every teaching tool has its advantages and disadvantages.

Interactive Response Simulations

Use

The interactive format that the computer provides is probably the best use because it allows students to participate in a way that no other medium provides. Thus, students can engage in a virtual laboratory experience with physical elements that are not subject to the human errors involved in doing such an experiment in the classroom. This is especially true in the physical sciences.

Another good use of the interactive format is in the form of simulation "games" with scientific themes. In these scenarios, students investigate a rain forest, or try to survive on an ocean journey. Students are able to input their responses to questions and get feedback as to the results of their success. Many simulations of this type are available in the marketplace.

Problems

Careful selection and screening is necessary to find the good examples. There are many good examples, but there are even more poor choices available. Typically these simulations use a lot of time. It needs to be time well spent, age appropriate etc.

Internet

Use

The internet is a valuable tool for research for both you and your students. Your students can use it to do projects, reports, and find answers to questions. You can use it to find lesson plan activities and good teaching ideas for units that need help.

Another use is for your class to communicate information and/or data with classes in other parts of the nation or world. There are prepared programs available that set up this communication in an organized way, allowing students to participate in collecting data for an experiment. You can also set up a sharing example on your own with another class. Units on weather and natural sciences lend themselves well to this kind of sharing. Children love to hear other children talk first hand about the places they live and your class can compare and contrast specific content areas.

Problems

Research is easy to accomplish on the internet, but be aware of the modern day problems. The source of the material may be unknown and may not be reliable. You need to assist students in verifying the source. Some teachers report more factual errors from using the internet than form using the "old-fashioned" books.

It is also easier for students to copy something and print it out without even re-phrasing or understanding the material. In the past, this copying could be done through tedious efforts, and that was also a problem, but sometimes something was absorbed during the copying stage.

Of course, if you are contacting other people, you again have to check out your sources. We have all heard about some of the horrors of computer contact misuse, and we must guard against these possibilities for the students in our care.

Prepared Programs

Use

Prepared programs will vary greatly in their quality and, therefore, careful selection is needed. Using a preview process, you are likely to find some excellent prepared programs that fit your needs for a particular situation. If the prepared program is not interactive, it is generally less useful, but it may still serve the purpose for a tutoring session, review, or presentation of a complex subject in one more format.

Problems

If the program is not interactive, it is not likely to be a substitute for good teaching. The cost and time spent often do not warrant the substitution for teacher presentation in an interesting fashion.

Programming Presentations

Use

If you have the equipment available and you are knowledgeable, the preparation of program presentations is an excellent experience for students. They can begin to learn the techniques of making slides, posters, etc.

Problems

The main problem is in the availability of equipment and the technical knowledge needed to teach these techniques. As always, you also have to balance the time spent against the learning that takes place.

General Problems

Using computers can become a management problem if not planned well. Frequently teachers do not have enough computers for the whole class and multiple lessons need to be prepared so that students can rotate in small groups to use the computers. The most important other general problem is the need for teacher training. If you are not up-to-date on the technology, take advantage of the many workshop opportunities being offered and add to your repertoire of good teaching skills.

Another general problem that we need to be aware of is the problem posed by the rapid rate of technological changes. Although students need to be familiar with today's technological tools, we need to teach for the future as well as the present. The technology of the future may be very different in form and function. Thus, teaching the skill should not be the only objective. Today's tools should be used in a problem-solving manner so that our students are prepared for the present and the future. When the technology changes, students who have learned to solve problems, will be able to adapt to the new era.

Technology Education

Technology has played a major role in civilization and students should get an introductory idea of its historic importance. They should also have some concept of designing and engineering construction. There are many wonderful design projects available in teaching resources from building bridges and towers to designing an alien. Often teachers use a design activity at the end of a unit so that students can synthesize what they have learned by developing the design project (an electronic circuit, for example). However, design activities can fit into any portion of your unit presentation. Design activities are part of the National Technology Standards.

Example Activity — Float it Down the River

Correlation to Technology Education

The example of technology education used here is an example in which students build a craft with a specific set of requirements. Many other examples can be used. This particular example allows students to engage in designing a craft to compete in a contest, and the contest adds an element of fun for the students. As described below, the activity opens up many avenues for exploration.

Introduction to Activity

"Float it Down the River" is an exciting design activity that involves students in a hands-on, creative activity in which they use higher order thinking skills in a highly motivational setting. Students are asked to design their own craft within a small set of guidelines. As they design their craft they are involved in inquiry science, they discover a number of scientific principles, they see the relationship between mathematics and science, and they apply mathematics in the process.

Description of Activity

The basic activity directs students to:

Construct a device using a minimum amount of materials that will carry the most weight down the river.

The materials provided for building include:

waxed paper, aluminum foil, plastic wrap, cellophane tape, and plastic drinking straws

The materials provided for working and testing include:

a clear plastic tub and a set of metric weights

Stated Conditions:

The device must float (not sink).

The size of all materials used must be measured in square centimeters.

The weight floated must be measured in grams.

The "contract" will be awarded to the team that builds the device using the least amount of materials to carry the greatest weight

(An appropriate student hand-out is included in this article.)

Lesson Analysis

As the students work to create their models they learn about density and buoyancy including many characteristics and features that affect floating in water. They try different shapes and discover the practicality and impracticality of various shapes. One group tried to work mathematically, for example, and tried to build a hexagon with the plastic wrap. They discovered that this was not a practical way to carry weights.

Obviously students discover the basics of floating devices and quickly learn that any water taken into the craft will cause the craft to sink (for example, through open straws). As they work, students also discover that distributing the weights in the craft affects the amount of weight the craft can support (small weights scattered throughout the craft work better than a large weight in the center).

The students are working with technology education and they are working with mathematics in a realistic setting. Ratios are used to determine the best craft according to the stated conditions, so students are interested in determining this for themselves. They also need to determine the area of each surface used in their craft before the ratio can even be used. The mathematics becomes an integral part of the activity.

This experiment can be used as an introduction, follow-up, or assessment for density, buoyancy or related concepts. The degree of detail and the breadth of concepts discussed can be modified according to the group, their grade level, and the interest exhibited.

The basic activity is included here but several modifications are also available for various groups, adding other dimensions to the activity. For example, one version sets a price on each item (e.g. waxed paper $.25 per square inch). With a specific list for all items used, students can determine the cost of their device, and the conditions of the contract could be changed to include this cost factor. Experimental devices can be included in the total cost, or might not be included. The difference allows for a lively discussion about the cost of research.

Another variation in the activity is the use of weights. Students can be allowed to test (do research) as they go along, or they might be asked to make the best craft without the use of weights and then test all devices at the end. Again, the value of research comes into the picture.

Simplifying the activity, instead of extending it as indicated above, is also possible. Students can be given no size requirement (except that the device fit into the container provided), and the size factor can become a part of the discussion. Although this reduces the amount of mathematics involved, it makes the activity easy for younger students. It also makes the activity more open-ended, allowing all conditions to become part of the discussion. There are numerous possibilities for adaptations. For all of the versions, the children find this to be an exciting activity and they are learning in an integrated open way.

Discussion of Correlation with Technology Education

Many other design activities are possible to develop using similar guidelines but based upon different subject areas. Many teachers have come up with design experiments of their own based upon the general concept. All of the design activities allow students to work cooperatively in learning mathematics, science, and technology in an exciting integrated format with open-ended possibilities. Design examples that are commonly used include designing bridges, towers, space suits, aliens, electric or magnetic games, etc. The sky is the limit once the idea of design is considered as an activity. In your university/college classroom, discuss this idea of technology education and develop designs that could be used in your elementary classroom.

Concluding Remarks

It is important to remember that technology education is included in the national standards. Children have fun with these design activities and they are easy to include in your curriculum. The emphasis on using educational technology sometimes obscures the aspect of technology education. Both are important parts of science education.

Selected Sample Websites

www.iste.org
International Society for Technology Education
Books, professional development, standards projects

www.infotoday.com
This site provides information on periodicals, conference materials, and products.

www.media-methods.com
This site provides information on resources, materials, and sample activities.

www.curriculumassociates.com
(This is an award-winning site with materials created by teachers for teachers.)

References

Brendzel, Sharon, Lucy Orfan and Robert Schuhmacher (2000). "Float it Down the River" *Science Scope* 24,2:12-14..

Ebenezer, Jazlin and Eddy Lau. (1999). *Science on the Internet.* Upper Saddle River, NJ: Prentice Hall.

International Technology Education Association. (2000). *Standards for Technological Literacy: Content for the Study of Technology*. Reston, VA: Technology for All American Children Project of International Technology Education Association

See Appendix F

Chapter 9
Adapting the Curriculum: Students With Special Needs

Introduction

Science lessons provide an excellent vehicle for children with special needs. Good science teaching strategies involve the use of hands-on experiments, demonstrations, models and visual materials. All of these strategies enhance learning for students with a variety of different needs. For example, vocabulary acquisition is fostered when students are provided with additional modes of presentation such as models or visual displays, along with the verbal presentation. This approach makes learning easier for bilingual students. In addition, concepts are more easily grasped. Research indicates this fact, and, common sense supports the idea. Having the hands-on materials also makes it easier for students with hearing difficulties, and students who have learning disabilities. For students with physical handicaps or visual impairment, science can easily be modified to accommodate them. This can be done with group work in which responsibilities are appropriately shared and with the opportunities for tactile senses to be used. In today's world with plastic equipment and other simple to use equipment, science is not a problem for special groups, and, science can even play a significant role in learning.

General Adaptations

The general adaptations that apply to all special groups include the use of group work, non-breakable and easy-to-use equipment, appropriate classroom settings, and the use of multiple presentation styles. In typical classroom situations, the various roles in the group are rotated among students, but this can be adapted to fit the needs of a special student. If the student has difficulty in manipulating objects, writing, speaking etc., another group member can be assigned that role. The student can rotate into the positions appropriate for him/her but does not need to be assigned a position that would cause a problem.

The use of simple non-breakable equipment and the easy current access to this equipment has been a boon to all elementary science teachers. It is of particular importance to some of the special needs students because awkward movements do not become safety hazards. Again, the appropriate classroom setting has always been something teachers should consider. Thus, a child with poor vision should be near the front of the room. This concept can be expanded with common sense to make room for wheelchairs, other needed devices, or simply providing enough space for the child to feel comfortable.

The use of various presentation styles is helpful to many children with different learning styles, and provides for better learning in general. Some specific steps may become essential for a child with special needs. For example, it may be necessary to have the directions in both written and oral form for learners with either hearing or visual impairment. Learning centers often allow you to set up different assignments for students to choose from. This concept will be mentioned again, specifically for some student groups.

Specific Adaptations for Special Needs

Physically Challenged

For the physically challenged, a teacher should use all of the ideas presented above with emphasis on reviewing the room setting and establishing group work to assist the individual. If the challenge is a physical one, there is no need to over simplify the material or use different material except when safety concerns might be the issue. If the physical handicap makes pouring difficult, one needs to assess the relative danger. Thus, if the solution being poured is water, there is no danger and the only need is for clean-up equipment to be readily available. If the solution is more caustic, a different group member can handle that situation. It is also, always essential that the child feel reasonably comfortable. If he/she prefers a different role to avoid embarrassment, for the most part, this should be respected (the exception might be when you have judged that the child is holding himself/herself back with unnecessary fears).

Visually Challenged

For the visually challenged, many of the general adaptations apply with some specific suggestions. Seating needs to be arranged for the child's best view. Directions and important information need to be oral, and may even need to be taped. If the visual difficulty is such that color-coding and/or brighter visuals would help, this should be arranged. When appropriate to the situation and safe to do so, the child should be allowed to explore the material by touching it and handling it.

Hearing Impaired

Accommodations for the hearing impaired child are in some senses the same and in some senses opposite those given for the visually impaired. Thus, directions need to be written as well as oral, but if the student can hear better by being near the front, this should be arranged. If small additions of sound will assist the child, a bell or chime, for example, these can be included. Since this child does rely on her/his other senses in a more profound way than the typical child, she/he may also want to use the sense of touch when appropriate.

Slow Learner

For the slow learner, multiple presentations work well. Directions and steps need to be clear, and the child may need some extra time to handle and review the equipment. Often, a learning center, available to all students, can include extra activities with the material presented differently so that the slow learner has several opportunities to examine the concepts.

Gifted

A child may be gifted in a variety of ways, so when possible allow him/her to use that gift (art, music etc.) in creative classroom situations. Working in groups often helps the child and prevents boredom while it helps his/her fellow classmates as the gifted child helps guide some of the questions etc.

Opportunities for enrichment should also be available. However, the teacher needs to be sensitive to the child's desire (a problem for some gifted children is the sense that they are always given extra work). The enrichment can be in the form of a special center, trip to the computer room and/or library when the child's interest encourages this, and/or special creative assignments. Often the best way to do this without singling the child out is to offer these enrichments for students who are interested and have the time. This allows all children to take advantage of the opportunities some of the time, and offers those who consistently finish early something interesting to do.

Example Activity — Cabbage Juice Indicator

Correlation with Special Needs

Most activities can be adapted for children with special needs. One needs to review the activity with the student in mind, consider the general and specific suggestions given in this chapter and use educational common sense to determine other specifics. If you decide that additional information is needed, consult the specialist in your district and/or appropriate research.

Introduction to Activity

Cabbage juice is a good homemade indicator of acids and bases. It turns colors in the presence of acids and bases.

The attached chart can be used to identify acids and bases using cabbage juice as the indicator.

Recipe for preparing cabbage juice to be used as an indicator.

Use chopped red cabbage (1 cup) and water (2 cups).

Boil until the purple color is obvious (avoid aluminum pots). This should only take a few minutes.

If you wish to preserve the solution for more than a few days, 1 ounce of isopropyl alcohol can be used (mark appropriately as poisonous). Refrigeration works well for a few days without alcohol and the prepared cabbage juice can be frozen for longer periods of time.

Description of Activity

Test your solution against known acids and bases. If there is a problem, start again.

Sample household substances to be used for testing:
lemon juice
cleanser
vinegar
dissolved baking soda
dissolved baking powder
hydrogen peroxide
milk of magnesia

Directions

- Use the cabbage juice to test your samples.
- Group the materials by color.
- Compare to the chart.
- What conclusions can you make?

Table 9.1 pH of Cabbage Juice

pH values

Strong Acid Neutral Strong Base

0——————————————————————————————14

Pink		White		Lavender		Purple		Blue		Green		Yellow

Lesson Analysis

Acids turn the cabbage juice shades of pink, bases turn the cabbage juice blue or green. These changes are not the changes that would occur if you mixed colors (as in paint colors). Cabbage juice acts as an example of an indicator. The conclusion will depend upon the background knowledge of your students. This can be an introductory activity. Students with no information about acids and bases can observe that there are two groups of substances, and this activity can be the springboard for introducing the types of substances (acids and bases).

If the students are already familiar with the concept of pH, the activity can be used as an example of application. Students can create a pH chart based upon their observations. You can even go one step further with the activity and use it as in inquiry investigation by comparing cabbage juice and spinach juice to determine whether either one is an indicator (only the cabbage juice is). This version of the activity can only be done if students are familiar with pH and can use other indicators to measure and draw conclusions.

In terms of application to this chapter, all of the standard advice applies. For safety reasons, only non-caustic household chemicals should be used and never allow students to have bleach and ammonia at the same time. Students can bring materials to test, but screen the materials. Use all plastic equipment for measurement and testing. Seat and assign students appropriately. Use oral and written directions including brightly colored examples of the chart. Use careful labeling of important items, and remind students to do the same. Have additional questions available for thought and investigation.

Discussion of Correlation with Special Needs

In your university/college classroom, discuss with your classmates and teacher any other suggestions or ideas for specific groups and choose one specific activity to adapt for one or more groups.

Concluding Remarks

Remember that children learn in different ways and a variety of teaching strategies is essential to good teaching. If a child in your class has a particular specific special need, this increases your need to adapt the curriculum and teaching to the specific situation. Use good teaching common sense and apply the most appropriate strategies. If you need additional assistance ask the "experts" in your school and/or do more reading on specific ideas for this type of adaptation.

Selected Sample Websites

www.cec.sped.org

The Council for Exceptional Children provides professional development information, publications, and information about standards. There is also a connection to ERIC clearinghouse providing information about lesson plans, research etc.

www.cal.org/crede
The Center for Applied Linguistics is a private, non-profit organization sponsoring a wide range of activities including research, teacher education etc. The Center for Resource in Education Diversity, and Excellence (CREDE) at the University of California at Santa Cruz provides research information, publications, contacts etc. for teachers.

www.gifted.uconn.edu
The University of Connecticut Center for Gifted Education and Talent Development connects to several sites. The National Research Center on the Gifted and Talented (NRC/GT) includes links to many educational resources and the link for Resources for Parents includes links to State Gifted Associations and Free US Department of Education Publications.

References

Armstrong, Thomas. (1994). *Multiple Intelligences in the Classroom.* Alexandria, VA: ASCD.

Brendzel, Sharon. (1995). "Hands-On Science for LEP Students" *NJTESOL-BE, Inc. Newsletter.*

Chapter 10
Relevant Research

Introduction

Many important traditional and current theorists are relevant for science teaching. This chapter includes a brief discussion of some of the most important.

Sample Theorists

John Dewey

John Dewey is a classical theorist whose ideas lay the groundwork for many of the modern constructivist views. He believed in learning by doing, very much the current philosophy and one that has always been applied to science teaching. Dewey emphasized the process of education in which the child reconstructs knowledge. The constructivist view employs a similar approach whereby the learner constructs his own knowledge. This approach is currently used in science education.

Jean Piaget

Piaget's work is by now classic and applied throughout the curriculum.

His basic stages apply to the learning development of students; both research and observation have supported his work. Most children in the elementary grades are in the Concrete Operational Stage where they need to see the experiment in order to grasp the consequences.

When they see something that behaves differently than expected, they need to adjust their view of the world. Piaget referred to this as *accommodation* and juxtaposed this against *assimilation* in which the behavior fits with the world-view and is more easily integrated into one's thinking.

Piaget's work on the development of conservation and the development of logical thinking helps us to understand the best way to present science. Since most children do not enter the Formal Operational stage until age 12, abstract reasoning should not be the focus of our science teaching. Even at that age, some hands-on concrete experiences facilitate the learning. Table 10.1, which follows, is a brief summary of Piaget's stages.

Table 10.1 Piaget's Stages of Cognitive Development

Stage	Age (in years)	Description
Sensorimotor	0-2	Objects lack permanence, reliance on senses
Pre-operational	2-7	Egocentric, magic, intuitive
Concrete Operational	7-12	Uses cause and effect, concrete experiences very useful, learns conservation
Formal Operational	12 and older	Can develop hypotheses, can use abstract thinking

Lev Vygotsky

Like Piaget, Vygotsky believed that learners construct their own knowledge, but his main concern was that of the role of social interaction in the learning process. He used the phrases *zone of proximal development* and *scaffolding* to explain his ideas. The *zone of proximal development* is used to indicate the ability of the child to learn with the help of others. Obviously, cooperative learning works well in this regard. If the teacher acts as the facilitator, the learning is maximized. The teacher acts as a *scaffold*, providing the right support to achieve the goal.

Jerome Bruner

Bruner is the father of discovery learning. His book *The Process of Education*, published in 1963 set the stage for discovery learning programs. The book summarized the work of scientists and educators meeting together at a conference in Wood's Hole during 1959. The meeting was for the purpose of suggesting approaches for the best ways to teach science.

Bruner stated "The schoolboy learning physics is a physicist."[1] In many ways, science teachers follow his ideas as our students perform investigations and conduct experiments. Bruner believed in inductive reasoning whereby the child develops the concepts from the problem solving experimental experience and generalizations leading to the overall concept.

He also believed in spiral education, the idea that the student can learn the concept in a limited way in the early grades and re-visit the ideas at a later grade. This enables the child to build on previous experience and knowledge, enriching his/her idea of the concept.

David Ausubel

Another theorist that many science teachers rely on is Ausubel. His work was more directed to verbal learning, and his suggestions of advanced organizers have been very helpful when presenting information. Basically, he suggested that we use *advanced organizers* to help connect the known to the unknown. He also emphasized organization, beginning with general ideas and moving to the specific, always connecting the new experiences to the old.

Although, as teachers in the classroom we often start with the specifics and use these experiences as building blocks, there are also many times when presenting verbal information, that Ausubel's ideas seem relevant. Ausubel believed in the deductive approach, from the generalization to the specific examples.

Howard Gardner

Howard Gardner's theory of multiple intelligences emphasizes the idea that students have different learning styles and they, therefore, learn best in different ways. One student may do very well with the traditional verbal learning styles, but another may need the visual approach to learn well. Gardner's list of intelli-

---

1 Bruner, *The Process of Education*, p. 14.

gences originally included seven and now number eight: linguistic (verbal), musical, spatial, logical-mathematical, bodily-kinesthetic, knowledge of self (interpersonal), knowledge of others (intrapersonal), and naturalist.

Gardner's ideas appeal to teachers, at least in the overall sense, because teachers see and experience the fact that children learn well in different ways. We do not, as teachers, have to determine the exact number of intelligences or their precise division, but we do need to provide different experiences. By doing this, we can help children utilize their best skills.

Benjamin Bloom

Bloom developed the idea of the Taxonomy of Cognitive Objectives. His work is classic and used by all educators to develop thinking skills. The basic idea is to help children use the higher order thinking skills, and not to restrict them to the lowest level skills. There are six levels, with knowledge being the lowest level and evaluation representing the highest thinking skill. Table 10.2 lists the stages with a brief definition.

Table 10.2 Bloom's Taxonomy

Level	Definition
6 Evaluation	Evaluate — judge, choose
5 Synthesis	Synthesize - bring ideas together, formu- late Ideas, design
4 Analysis	Analyze — formulate hypotheses, explain causes
3 Application	Apply — solve a similar problem or use the idea in a similar setting
2 Comprehension	Comprehend – understand, explain
1 Knowledge	Know — define, recall

Cooperative Learning

Cooperative learning has become a standard classroom approach in recent years. There is good reason for this. Children working in groups have the chance to talk to each other and learn from each other. During group learning, the number of interactions for each student increases many times over the old approach with the teacher at the front, calling on one student at a time. The groundbreaking work for this cooperative learning approach was done by Johnson and Johnson and R. Slavin in the 1970's. Since that time, there has been a great deal of research to support the cooperative learning theories.

Some of the most important ideas taken from the original research include the concepts of both group and individual responsibility, mixed ability grouping and task assignment. Work proceeds best when the group members feel responsible for their own work and for that of their group members. Mixed-ability groupings allow students to learn from each other, and task assignment helps control the management issues.

Often, in the classroom the group work does not follow all of the ideas of the originators, but, instead, teachers modify the approach to fit their needs and the needs of their students. The group learning structure still works well and gives the students a sense of involvement.

Testing Implications

In 1988, the US participated in the first International Assessment of Educational Progress (IAEP). The country was surprised to find out that our eighth grade students did not fare well compared to students in other countries. The tests were repeated in 1992, 1995 and 1999. In 1995 the test was called TIMSS (Third International Mathematics and Science Study) and in 1999 it was labeled TIMSS-R (TIMSS –Repeat).

For a summary comparison of the results see Tables 3 and 4.

The results of the tests indicated that our eighth grade students were near the bottom of the lists of ranked countries in 1988 and 1992. By 1995, our scores were in the middle group range, but well below the desired performance for the US. The results of the testing for the year 1999 indicate approximately the same position.

Our fourth grade youngsters do perform well compared to their peers, but the comparison drops off in grade eight and grade twelve. The tests use questions that require more than recall, more than lower level knowledge, and our students do not appear to be prepared for this challenge.

There are many sub arguments and discussions. Individual schools have challenged the data and have tested their students to show that their scores are comparable to the higher achievements. Some educators have pointed out that our top students do compete favorably. There have also been arguments as to student selection in different countries.

However, the bottom line is still that our average students do not compare well with the average students form many other countries and the number of students performing in the top is a lower percentage than the top performing countries. For example, in 1999, 15% of US students achieved the highest level while Korea had 22% reaching this level. Obviously this poses concerns for everyone in teaching and generally for the community.

An examination of Tables 10.3 and 10.4 reveals that US scores are close to the average international scores, but our placement is low in comparison to other countries in grade eight (the results are worse in grade twelve). As noted earlier, our fourth grade students do compare favorably. What happens between fourth and eighth grade? Curriculum analyses, content analyses and other forms of analyses have been made. The most frequent criticism that has been levied is that we attempt to cover too much material without the depth of problem solving skills needed.

Table 10. 3 Comparison of International Tests from 1988 –1999, Eighth Grade

Test /Year	US Score	Int. Avg.	Int. Avg. %	% Correct US	Placement Position
IAEP/1988	478.5	*505	*60.7	*58.2	9th out of 12
IAEP/1992	NA	NA	*69.5	67	13th out of 15
TIMSS/1995	534	516	66	66	~14-15 out of 23
TIMSS-R/1999	515	488	67	67	18 out of 35

Notes:

Test Names
IAEP refers to International Assessment
TIMSS refers to Third International Mathematics and Science Study (the name given to the third test) and of Educational Progress (the name given to the first two tests)
TIMSS-R is TIMSS –Repeat

* Calculated numbers based on data available in the original documents. The % figures for 1988 were actually difficult to calculate because they were based on bar graph readings.

~ 14-15 because the US and Hong Kong had the same score, occupying spaces 14 and 15

NA = Not available

Table 10. 4 Comparison of International Tests from 1988 - 1999, Fourth Grade

Test /Year	US Score	Int. Avg.	Int. Avg. %	% Cor-rect US	Placement Posi-tion
IAEP/1988	NA				
IAEP/1992	NA	NA	NA	65	3 out of 10 (Comprehensive countries)
TIMSS/1995	642	514	NA	NA	3 out of 17
TIMSS-R/1999	NA				

Notes:
The Fourth grade was not tested in 1988 and 1999.

Test Names

IAEP refers to International Assessment of Educational Progress (the name given to the first two tests)
TIMSS refers to Third International Mathematics and Science Study (the name given to the third test) and
TIMSS-R is TIMSS –Repeat

The thrust to teach according to the standards has been in place as long as these tests have been in place. *The National Science Standards* were published in 1996 with earlier drafts and workshops. Perhaps the ideas have not yet reached all of the teaching population, and of course, it takes time for students to adapt to the new ideas and perform better on tests. It still leaves educators questioning why we are not seeing more results. This trend will be watched carefully in the future. Our results will continue to be compared to those whose results are better. Our methods will continue to be analyzed, and we may even see some revision of our current approaches.

Problem solving is at the core. Currently, we strongly favor student group participation. Japan, on the other hand, uses teacher directed problem solving (See Chapter 1). Would it work as well with our students? Cultural differences lead us to believe that it would not. However, there is little question that problem solving must be an essential part of our teaching. Unfortunately, there is still evidence that this approach is not being used throughout the country.

Since we cannot afford to accept the status quo, we have to keep analyzing our results and adapting our techniques, if need be. Teaching is not a static profession. It never was, and we will need to continue to be on the alert.

Concluding Remarks

Of course, no single chapter summary of research can be complete. The author determines the choice and there are so many appropriate possibilities. With this in mind, the summaries in this chapter can be viewed as good examples. A survey of other texts on the same topic would indicate that there are a few examples in nearly every text (Piaget, for example) with other choices varying from author to author. Thus, the idea here is to expose the reader to at least a few of the most commonly used, most appropriate research examples.

However, it is important to note that research is relevant and has helped teachers to do a better job. On your own you should explore more research, especially in areas that seem the most relevant to you as a teacher. If a new idea is presented, be open-minded and try it. If it works, great, add it to your repertoire. If it doesn't work, don't give up immediately. Try to find out why it worked for others and explore some possibilities before giving up.

No new lesson plan is presented with this chapter. Instead, in your university/college classroom, discuss the research presented here (or other research) and give examples of classroom applications.

Selected Sample Websites

www.eric.ed.gov

The ERIC Clearinghouse has been a long-time standard research system. You can obtain information through Selected Fields (using authors, keywords, titles etc.) or through the Eric Thesaurus (select your topic). There is also a Comparative Chart of Eric Special Access Points (clicking on this also allows you to see the list of Clearinghouses. Once you select an area and supply the appropriate information, you submit your request and you will obtain a long list of documents. You can then select a particular document and read the abstract online. The abstract sheet will also provide information about how to obtain the full document if you are interested. Many of them are available through the Eric Document Reproduction Service for a fee.

Use a standard search engine and list the research item you are interested in. You may get information and/or latest publications. Yahoo actually provides a document search service and allows you to obtain the document for a fee. Check with other standard search engines that you use.

www.eric.ed.gov
Connection to the Eric Clearinghouse list of research and resources with many additional links including a connection to the ERIC Clearinghouse link for Elementary and Early Childhood Education.

www.wisc.edu/
This site leads to several listings for national organizations in science education. The Coalition for Educators in the Life Sciences provides a listing under the site

www.wisc.edu/cel and another site lists Professional Societies in Higher Education and Science Education

(www.wiscedu/cbe/cel/monograph/mono6X1.htm)
This site can be reached through the opening site listing of **www.wisc.edu/.**

nces.ed.gov is the web site of the National Center for Education Statistics. Many materials are available including the report on the findings of TIMSS –R which is currently available only online, but will be published shortly.

References

Armstrong, Thomas. (1994). *Multiple Intelligences in the Classroom,* Alexandria, VA: ASCD.

Ausubel, D.P. (1963). *The Psychology of Meaningful Verbal Learning.* New York: Grune and Stratton.

Bloom, B.S. (1956). *Taxonomy of Educational Objectives. Handbook I. Cognitive Domain.* New York: David McKay.

Bruner, J. S. (1963). *The Process of Education.* New York: Vintage Books.

Dewey, John. (1916, Paperback ed., 1961). *Democracy and Education.* New York: Macmillan.

The International Assessment of Educational Progress. (1992). *Learning Science.* Princeton, New Jersey: Educational Testing Service.

Johnson, D.W. and R. Johnson. (1975). *Learning Together and Alone.* Englewood Cliffs, NJ: Prentice-Hall.

Lapointe, Archie E. et al. (1989). *A World of Differences: An International Assessment_of Mathematics and Science*. Princeton, New Jersey: Educational Testing Service.

National Center for Educational Statistics (2000). *Pursuing Excellence: Comparisons of International Eighth-Grade Mathematics and Science Achievement from a U.S. Perspective, 1995 and 1999.* Jessup, MD: US Department of Education (National Center for Education Statistics).

Piaget, Jean and Barbel Inhelder (1969). *The Psychology of the Child.* New York: Basic Books.

Schmidt, William H., et.al. (1997). *A Splintered Vision: An Investigation of US Science and Mathematics Education.* Boston, Kluwer Academic Publishers.

TIMSS. (1997). *Third International Mathematics and Science Study*. Philadelphia, PA:

Mid-Atlantic Eisenhower Consortium for Mathematics and Science Education.

Vygotsky, L. (1978). *Mind in Society*. Cambridge, MA: Harvard University Press.

Chapter 11
Resources

Introduction

The resources available for teaching science today are abundant; time is our only limitation in gaining access to the wealth of materials available. For each resource discussed in this chapter, a brief description will be given, including appropriate use in the classroom and a discussion of some of the limitations. In addition to the sample lesson at the end of the chapter, examples of lessons are provided for a few of the resources.

Resource Examples

Text Packages

In science classes the text should be a back-up to the activity, providing a helpful explanation of the concepts to assist students in synthesizing material. However, today's text packages can be very helpful to the teacher. The new texts themselves usually have wonderful photographs and charts that can be used for visual representations and to help motivate students. The activities that are used are often easy to do and the chapters generally include some activities.

In addition to the actual text and teacher text, there are a variety of workbook packages, assessment tools, and many other helpful sections. Some of these special materials provided in the text packages include overheads, correlation strategies with other subjects, matrices that correlate units with standards, ideas for projects etc. It can actually take the teacher more than one year to sort through all of the offerings and select and utilize them well. This, of course, is keeping in mind that other subjects are being taught and that adjustment to the new text and materials involves time in and of itself.

The major disadvantage of textbooks is that teachers often let the text serve as the driving force in setting up the curriculum. This removes some of the teacher's own creativity and, of course, some text materials are better than others. Students then receive the text package instead of the teacher's best efforts. Moreover, even the best texts often have activities in the recipe format and teachers should revise many of these activities. The assessment portions are becoming more in line with alternative assessments but may still be uneven. For the best teaching, each text needs to be evaluated and adjusted as needed.

Kits

Many companies make science kits and packages of activities. Some of these are made to accompany particular texts and others are organized for a specific unit of study. The advantage to the teacher is that a classroom set of materials is pre-packaged and set up for a specific experiment or activity. The prepared "teacher guides" give directions and ideas to run the activity and sometimes they are organized in an open-ended way. Kits are available form supply houses, text publishers and organizations. Two currently popular sets are FOSS and STC. Although there are many kits (See Appendix A), FOSS and STC have been chosen to serve as examples of the current kits. A brief review of the older kits is also included.

FOSS

FOSS stands for Full Option Science System. It was developed at the university of California at Berkeley. The developers have indicated that these kits can be used in conjunction with other materials or as the basic science curriculum in your school. There are several modules for each grade level and a school that purchases the entire set will be able to build a curriculum around the materials.

On the other hand, a school can purchase a particular module to assist a teacher in presenting a topic. If this is done, care must be taken to examine the kit and determine what materials are needed in addition to those provided in the kit. There are usually some basic materials that are supplied if the whole set is purchased but are not packaged with the individual module purchased. These basic materials may already exist in your supply cabinet because they include items such as beakers, spoons etc., but, if needed, they can also be purchased. It is also important to note that some of the materials are consumable and these need to be replenished each year. The activities cover examples from life science, physical science and earth science.

Teachers who use the materials as a complete curriculum say that FOSS does provide good activities, but these teachers also feel that it is difficult to conduct good lessons without any written material. They find themselves spending a lot of time copying materials and finding reading materials to back up the activity presentations. Often, this is a problem with administrators who purchase a package and want this to be **the** curriculum. A note of caution for administrators and teachers, a single source does not provide a good curriculum. Whether it is a text package, a kit package or the laser disc, one type of material will not do the whole job. Teachers need a variety of materials. Activities need to be at the front and center but reading materials and visual aids are needed to back-up the activity work.

Another comment from teachers using FOSS is that the activities are not as open as they could be. Many teachers have found ways to open the activities more than they are currently presented. In summary, these materials can be very useful in the classroom. The FOSS materials make activity presentation less cumbersome than collecting everything individually.

This activity from AIMS has the students making measurements in the metric system with simulated activities from the Olympics. The four activities are: Paper Straw Javelin Throw, Paper Plate Discus, Cotton Ball Shot Put, and Right-Handed Marble Grab. In each activity the student estimates his/her ability and then actually make measurements. The students work in groups so they can obtain several records and also well planned and form a good basis for providing investigations and these activities are usually more open than most standard text activities.

STC

STC stands for Science and Technology for Children. The materials are prepared for grades 1-6 and the focus is on the application of technology to science. As with FOSS, materials are available for physical science, life science and earth science. The set up is similar to that of FOSS in that a teacher guide is available, materials are packaged for activities, and some materials are consumable. Many of the comments about FOSS apply to STC. The main difference is the orientation of the subject matter. STC does incorporate more technology. If you are allowed to select the kits, review the literature and some sample units to determine which kit fits your particular curriculum needs best.

SCIS, SAPA, ESS

There are several other kits of materials prepared by different groups including SCIS (Science Curriculum Improvement Study), SAPA (Science, A Process Approach) and ESS (Elementary Science Study). These three were first developed in the sixties and they all include kits, teachers guides and experimental set ups. Although each of these is activity centered, the approach and organization is different. Both SCIS and SAPA have been updated and are still in use. To get more information about these curriculum programs, see Appendix A for the addresses.

Another suggestion is for you, as the teacher, to check the storerooms of the schools in which you work. Often, some of the older versions of these kits are sitting collecting dust, and these can provide excellent activities for classroom use. This is especially important in districts where equipment budgets are small. You may find a wealth of materials in the supply room. Even without the manuals, some of these materials can be put to good use, and even if your school does not have the budget to purchase a whole new series, it may be able to purchase the needed accompanying manuals or a few additional materials to make these programs work for you.

Curriculum Series

Some of the curriculum programs are basically sourcebooks of activities. The two most popular are AIMS and GEMS.

AIMS (Activities Integrating Mathematics and Science)

These activities are fun for the students to do. Once purchased, the rights to reproduce class copies are included. The AIMS Education Foundation conducts workshops, produces newsletters and has a website. There are several books organized by grade level and by general theme and you can leaf though the pages to find appropriate activities for your current topic. The materials do not come with the book, but you can purchase them from the AIMS Foundation or another general supply house (See Appendix F).

*Activity Example – Metric Olympics (AIMS)*calculate averages (if desired). The Javelin Throw, Discus, and Shot Put are measured in centimeters and the Marble Grab is estimated and measured in numbers. Thus students gain familiarity with both linear measurement and numbers.

For safety reason, a suggested modification to the "Marble Grab" is to use beans instead. This is simply a matter of safety. If the marbles fall and roll, a student could slip on one. Also, the marbles do make a lot more noise when they fall, but this is less important than the safety issue. The beans do work equally well.

GEMS (Great Explorations in Math and Science)

GEMS is also a series of books with good activities integrating science and mathematics. The open-ended activities fit well with the national standards and have therefore been recommended by groups working with the standards. The topics fit well into a number of curriculum units and give the teacher flexibility. However, teachers do need to collect the materials themselves.

Special Books and Projects

Several organizations have published theme-centered activity resource books. Some of these cannot be purchased unless you attend the workshop, where the materials are explained and sample activities are tried. It is well worth the time spent (usually one day) to obtain these wonderful resources. Some of the best include Project WILD, Project WET, Project Learning Tree, BRIDGES, etc. (See Appendix B)

Project WILD and Project Aquatic WILD

These programs were developed jointly by the Council of Environmental Education and Western Association of Fish and Wildlife Agencies. The programs are available in all 50 states, the District of Columbia, Puerto Rico and several foreign countries. In most states these programs are available through the state wildlife agency or the department of education. The activities include simulations and games, presenting the concepts through activities that are fun for the students. Activities are adaptable to grades K-12 with guidelines indicating the most appropriate grade levels. The pages are easily reproducible and most of the activities do not require any special materials. Usually, the materials are simply paper, pencil, rulers and other simple tools using the reproducible pages in the book. Many of our university/college students have been introduced to these workshops and have loved the activities available.

Project WET

This project was created with funding supplied by the US Department of the Interior's Bureau of Reclamation and by the Western Regional Environmental Education Council. It is now available through *Project WET* coordinators in 46 states. The activities are similar to those of *Project WILD and Aquatic WILD.* The activities are simulations and games using everyday materials and reproducible pages. The suggested grade levels are indicated and the overall range is also K-12. Like *Project WILD*, the programs have been highly successful and both of these projects are well worth the day's workshop to gain the experience and receive the materials.

Activity Example Project WET

One of my favorite activities from *Project WET* is an activity on the water cycle entitled "The Incredible Journey." The activity provides for students to simulate various paths a water droplet can take. With this simulation, students become the water molecules moving through the different pathways. The places water can move through are categorized into nine stations: Clouds, Plants, Animals, Rivers, Oceans, Lakes, Ground Water, Soil, and Glaciers. The *Project Wet* sourcebook sets this up as a game in which the roll of the die determines where water will go (the labels on the sides of the die represent the options for pathways water can follow). Details of the patterns to follow are provided in the description of the activity and the appropriate tables are also provided.

The NJ division of *Project WET* has modified this activity and my students enjoy this modification. In this version, each station is set up around the room in envelopes (or boxes). Inside each envelope is a list of possible pathways. Each student receives a starting point station (listed above) and proceeds to that station to pick up directions. He/she then takes a slip from the envelope and follows that direction, returning the paper to the envelope. Any number of pathways can occur. Students should keep a record of their pathways for discussion at the end of the activity. The way the activity has been set up children need to understand what the direction indicates, requiring simple thinking. Two examples will illustrate that this is relatively easy: "Water leaves the plant through the process of transpiration" means that the next step is the Cloud station and "Water remains in the current of the river" means that the next step is for the student droplet to stay in the river. However, if you do this activity with younger children, you can follow the directions given in the original version or include this notation on the slip of paper in the envelope.

Whatever version you use, this is a fun activity and the wrap up discussion gives students a much better understanding of the water cycle than most traditional presentations.

Project Learning Tree

The *Project Learning Tree Environmental Activity Guide* was prepared by the American Forest Industry for grades K-8. The guide provides a lesson plan related to environmental themes. The book is given out at workshops run by state agencies.

WOW

The *Wonders of Wetlands Guide* distributed by the Project WET organizations listed above. (Incidentally, the same acronym is used by the World Wildlife Fund for its magazine-style primer on biodiversity, prepared for use with middle school children.) As the name implies, these activities center around water and wetlands, and the activities are similar in format to those found in Project WET.

State Projects

Check with your Department of Education and the various science related agencies in your state to see what special projects have been developed. The NJ Audubon Society, for example has developed *Bridges to the Natural World.* This project is prepared for grades K-6 and has activities centered around habitats. There is also a content section on habitats and a section on outdoor activities.

Activity Books

In many ways the activity books are like the special projects because each activity book includes a large number of activities that you can use or adapt for your classroom, according to the topic. Each author chooses the organization, but, for the most part, the organization is around a content theme. Many of these activities are in standard recipe format, but they are still useful and can often be adapted. Most of them use materials that are simple to find around the house and most of the activities have been field-tested. The variety and availability is extensive. Glance through them and select what you need. The best way to do this is probably to visit a resource center that has several books. You do not need to use every activity, just select the ones that fit your lesson plans. Another way to do this is to request that your school buy a collection for teachers to share. Setting up a teacher resource section in the library or teacher's room is an excellent way to have many sources available with a minimum cost.

Texts for courses similar to this one also contain activities, often in a separate section or even a separate book. Some of these should be available in a college or university library. The college/university library may also have a resource section of student texts. Other student texts that are different from the one your district is using may have different activities that you can explore and add to your collection. If your district is planning to purchase a new text, companies will provide samples for review.

Appendix B includes an activity book bibliography and the last appendix is a regular bibliography with the textbooks included. Appendix C is a list of organizations that provide materials for teaching.

Conferences

One of the best ways to stay up-to-date and get additional ideas in a short time is through workshops and conferences. Your principal or science coordinator should get many materials advertising these sessions. If you are not getting the information ask, so that the papers do not find their way to the circular file.

Many college/universities and various state agencies offer these workshops. Regional and state branches of the National Science Teacher's Association also offer workshops and many other related organizations run programs. Most of these are worth the time, but check with colleagues, friends and faculty to determine the reputation and decide which ones to attend. Check the cost. Most are reasonable, but there are some that you may not want to pay for unless your district is willing to commit the funds. The appendices list the various resources discussed in this chapter. Many of these organizations provide workshops as well as materials.

Field Trips

Field trips are an excellent way to augment science teaching. Local or regional museums, and natural sites offer a variety of possibilities. Most of these places will send you the literature and let you know about the special school programs available. If the budget fits your school situation and the trip is relevant, it can be a wonderful learning experience.

A few helpful hints should be useful. It is essential that you give the students a focus while they are on the trip so that they will be able to learn something and not just play. A worksheet is often the simplest way to accomplish this goal. Without guidance, the children might not focus, but the worksheet should not be so laborious as to prevent them from enjoying the activity. Look for a balance. In order to do this, you should preview the field trip and lay out your plan.

Also, know your school and try to select places that the students do not frequently visit on their own. Sometimes, smaller, local museums prepare excellent programs and these may not be the ones the parents take the children to visit. You may also want to offer a list of possibilities for interested parents.

If the budget does not permit a field trip that goes beyond the school, there are some outdoor activities that can serve as field trips in the appropriate units. Examine your surrounding resources and learn to make the most of them. Check with community services, both civil and retail. Often there is a relevant activity in your own neighborhood.

Outside Speakers

Like field trips, this type of activity can add breath and variety to your classroom. Many organizations offer a speakers bureau including conservation agencies, energy suppliers, museums etc. Again, you need to check your local resources and see what is available. Ask your colleagues and share notes. If no one else has tried it, give it a try. If the speaker is not what you want, you will not add it to your list, but you may be very happy with the results.

Many of these organizations offer these services free or for nominal fees, but check, because some may not be in your school budget. Some of these organizations also offer traveling displays/supplies. These can be borrowed for a period of time and can be used as visual supplements in your class. (Appendix C is a list of organizations.)

Don't forget to consider your parent pool as a possible resource, but be prepared to help set up the session in a way that is flexible (not all parents know how to speak to children as a group).

Magazines

Magazines and newsletters provide an endless source of materials for the science class. There are professional magazines and newsletters for teachers that have excellent activities and there are magazines and newsletters prepared especially for children. Sometimes schools are willing to purchase copies for the library or a set of newsletters for each grade level so that teachers can share. See Appendix E for a list and consider asking your school to purchase some as shared resource materials. In addition to the magazines prepared specifically for science teachers, most professional education magazines that are general in scope include some science ideas. (Appendix D is a list of periodicals and magazines for both teachers and students.)

Resource Centers

These vary from state to state but most states have educational resources centers. Sometimes they are separate agencies. Sometimes they are connected to the college/university system and sometimes they are connected to museums or agencies. Again, I suggest you ask around and find these resources. These Resource Centers usually have some of the resources discussed in this chapter available to you. Some lending audio visual libraries are also available.

Technology

Technology offers many resources. See Chapter 8 for a discussion of these resources and see Appendix F.

Activity – School Yard Erosion and Terrain Studies[1]

Correlation with Resources

A few lesson examples have been mentioned throughout this chapter. As in most other chapters, one full sample lesion is provided. For this chapter, the example chosen is a sample lesson from a magazine, *Science Scope*.

Introduction to Activity

Erosion is a natural process by which land is worn away by water, wind and ice. The most common agent is water. As it flows over land, water carries particles of earth away with it and deposits them elsewhere along its path. Such erosion is often easily seen in coastal areas, where noticeable amounts of land can be lost each year, and along rivers and streams. Runoff is also responsible for a significant amount of erosion.

Although erosion by flowing water is a fact of life, several factors affect the amount of erosion that occurs. These include the volume of the water, it's velocity, and the types of terrain over which it flows.

1Brendzel, Sharon (1994) School Yard Erosion and Terrain Studies. *Science Scope.*

Erosion occurs all around us, but most students are generally not aware of the damage that erosion can cause. This damage can be reduced if consequences that result from man's building and farming activities are considered. Students need to understand the control that man does have in order to be well informed citizens.

Some simple outdoor activities can be used to simulate the process of erosion and show how various factors affect the amount of erosion that occurs. The first two activities involve pouring water over miniature hills of sand and soil. These can be conducted as teacher demonstrations, or if sufficient space and materials are available, by students working in small groups. These activities work best when done outside, and the instructions are given for outside activities. However, these first two activities can definitely be modified for indoor use.

Description of Activity

Erosion factors

To demonstrate the relationship between volume and erosion, take the class outdoors and make two hills about 50cm high. The hills should be the same shape and composition and close enough to simultaneously pour water on them. Fill two identical pour bottles (for example, soda bottles) with substantially different volumes of water (one-third liter and one liter can be used). Pour the water over the hills at approximately the same rate.

After the demonstration, ask the class questions such as

- Which hill eroded more?
- Which hill eroded most quickly?
- What was different about the cause of the erosion?
- What can you conclude from this experiment?

Next, demonstrate the effect of velocity on erosion. Make two new hills and fill two containers with different-sized openings with the same volume of water. A soda bottle and a squeeze bottle from liquid soap are different enough to be used to illustrate volume differences. Pour the water over the hills and ask the class what they observe. The water from the *soda bottle* flows at a greater velocity and therefore causes more erosion. Observations are mostly qualitative, but an estimate of the quantitative differences can also be made. Question the class about their observations.

Types of terrain

Once students understand how volume and velocity affect the rate of erosion, they can go on to study how the type of terrain relates to runoff rate, and therefore the amount of erosion that occurs. To compare runoff rates for various types of terrain, students work in groups of four and use soda bottles to pour water over terrain such as areas covered with vegetation, bare soil, and concrete or asphalt. These three types of terrain are easily found around most schools.

From this simple experiment, students should be able to observe the importance of vegetation in preventing runoff and erosion. Water seeps into the ground relatively quickly when there is vegetation on the ground, and therefore does not flow rapidly over the land and wash away noticeable amounts of soil.

Water poured on bare soil spreads over a wider area and is likely to wash away some of the soil (depending on how packed and dry the soil is). Packed soil erodes more slowly than loose soil but erodes more quickly than soil held in place by the roots of plants. No noticeable absorption takes place in areas paved with concrete or asphalt, and runoff can cause erosion in the soil around the paved area.

It is important to remember that differences in results will be observed depending upon the relative moisture of the soil prior to the experiment. If time permits, students can repeat the activity on several different days and compare their results. Graphs of the results can further extend the activity.

Materials for Terrain Study

(For students working in groups of four)

- Two plastic soda bottles of the same size
- Timer with a second indicator
- Measuring tape
- Notebook and pencil for recording results
- Source of water such as a hose or pail of water

Procedure

1. Look around the school grounds and locate the following three types of terrain: an area covered by vegetation, an area of bare soil, and an area paved with concrete or asphalt.
2. Place a mark near the top of *each* soda bottle and fill to the mark. Beginning with any area you prefer, pour the water over the terrain turning the soda bottle upside down and holding it approximately 30cm from the ground.
3. Note what happens to the water. Record the time it takes for the water to seep into the ground. How much is absorbed, and how much runoff occurs (estimate percentages)? With a measuring tape, measure the approximate perimeter of the area affected by runoff. How much erosion occurs as a result of runoff? Record any other observations.
4. Conduct two more trials for this type of terrain in different locations. Be sure to keep the volume and velocity the same so that the rate and extent of runoff will reflect the terrain itself, rather than some other factor.
5. Continue for the two other types of terrain and compare the results. What generalizations can you make about the three types of terrain? How does the type of terrain relate to the amount of runoff and the amount of erosion?

Questions

1. What happens when water is poured over the land covered with vegetation?
2. What happens to water absorption with bare soil?
3. How much erosion occurs when water flows over land without vegetation?
4. What happens to water poured onto concrete or asphalt?
5. What do you think happens when a large portion of an area is paved, for example for a highway?

Answers

1. Most of it seeps into the ground and some runoff occurs, depending upon the volume of water and the amount of vegetation.
2. The water seeps into the ground more slowly, and the runoff covers a larger area. However, this will vary with the moisture content and composition of the soil. If there has been extensive rain and the ground is saturated, the water may form a puddle. If the ground is completely dry, absorption may be slow initially. In addition, clay soils will absorb water more slowly than sandy soils.
3. Since the soil is not held in place, it is washed away.
4. It cannot be absorbed and it flows away almost as quickly as it is poured.
5. The water flows over the area and is carried away without absorbing into the ground. The water then flows over the adjacent soil with a volume and speed that causes extensive soil erosion.

Follow-up discussion

After the activity, draw connections between students' results and real-world concerns about land use. Discuss the problems that can arise when too much of an area is paved over. Discuss the significant problems caused by erosion and ways that individuals and governments try to control it. Students may even be able to find and discuss local news articles relating to the problems of maintaining open spaces. Ask the class whether the government should have any control over the use of privately owned lands, and if so, to what extent.

Students should understand that erosion varies due to natural conditions and man-made conditions. Although man cannot control the changes due to natural conditions, we need to be aware of the changes we can control. For example, in areas of heavy rainfall, large areas of vegetation are needed to help absorb the water. Building plans should be made with this in mind. The interaction between natural consequences and poor planning by man can lead to heavy floods which result in loss of property and soil. Studies of erosion should lead to an understanding of the factors affecting erosion rates and the need for man to consider erosion rates in planning buildings, farms, and communities.

<u>Lesson Analysis</u>

This lesson example actually covered two different lessons related to erosion (erosion factors, and types of terrain). Therefore, the discussion, relating to the content and concepts of each activity, was discussed within the activity.

Within this section, it is deemed worthwhile to make a general comment about earth science study, because earth science has often been left out of the elementary classroom curriculum. It is, however, important and useful to teach earth science in the elementary classroom. Students should have some exposure to all content areas of science and, additionally, earth science studies help build an understanding of our environment and many of the problems facing our society. Fortunately, there are many simple and effective techniques that can be used to teach earth science as indicated in the erosion examples illustrated above.

Discussion of Correlation with Resources

This lesson example is one of many choices. In addition, other examples were briefly described as part of the appropriate sections. If possible, in your university/college classroom, examine some of the resources described in this chapter and choose a few good lessons to discuss with your classmates. Describe your view of the value of the resource reviewed by your group and describe the lessons you have selected so that you can share them with your classmates.

Concluding Remarks

As stated earlier the resources for teaching science are extensive. Whole books on particular resources are available and comprehensive resource guides are also available in book form. This chapter summary gives an overview of the types of materials available and briefly describes how to find some of these resources. In order to personalize this to your community and area you will need to investigate the local resources. You should also examine the various appendices, subdivided by topic for listings of some of the most popular examples of given resources.

Sample Selected Websites

www.thegateway.org
The gateway to Educational Materials provides access to information on a large number of topics. You can select from broad topics, and narrower topics (for example, Science s a broad topic and Geology as a narrow topic). Within each of the topics you will find information on research projects, lesson plans, curriculum units and other relevant information.

www.muohio.edu/Dragonfly/index.htmlx
The site has pages with information on the current topic in the magazine. You do not need the magazine to review this information.

pao.gsfc.nasa.gov/gsfc/educ/k-12/k-8/k-8.htm
This includes a large number of activities on a wide range of subjects prepared by Goddard Space Center.

www.eric.ed.gov
Connection to the Eric Clearinghouse list of research and resources with many additional links including a connection to the ERIC Clearinghouse link for Elementary and Early Childhood Education.

www.wisc.edu/
This site leads to several listings for national organizations in science education. The Coalition for Educators in the Life Sciences provides a listing under the site.

www.wisc.edu/cel and another site lists Professional Societies in Higher Education and Science Education

(www.wiscedu/cbe/cel/monograph/mono6X1.htm)
This site can be reached through the opening site listing of **www.wisc.edu/.**

References

AIMS (Activities for Integrating Mathematics and Science). AIMS Education Foundation, PO Box 8120, Fresno, CA 93747-8120.

Brendzel, Sharon. (1994). "School Yard Erosion and Terrain Studies" *Science Scope* 17,7:36-38.

Project Wet. (1995). *Curriculum and Activity Guide.* Houston, Texas: Council for Environmental Education.

Also See Appendices Listed Below and Bibliography

Appendix A	Curriculum Kits
Appendix B	Activity Books and Projects
Appendix C	Organizations
Appendix D	Periodicals and Newsletters
Appendix E	Suppliers
Appendix F	Technology

List of Appendices

Appendix A Curriculum Kits

FOSS (Full Option Science System)
K-6 Curriculum, 27 modules, all science areas
Developed by Lawrence Hall of Science, Berkeley, CA. 1993
Available: Delta Education, Hudson, N.H.

Insights (Improving Urban Elementary Science)
K-6 Curriculum, 17 modules, all sciences
Developed by Education Development Center, Newton MA,.1997
Available: Dubuque, Iowa: Kendall/Hunt

SAPA (Science – A Process Approach)
K-6 Curriculum, process oriented
Sponsored by the AAAS (American Association for the Advancement of Science), 1962
Available: Delta Education, Hudson, NH

SCIS (Science Curriculum Improvement Study)
K-6 science processes, current series has 13 units organized into physical-earth science, and life-environmental science categories.
Developed originally by Lawrence Hall of Science, Berkeley CA. Several up-dates have been developed. The current version is SCIS3+
Available: Delta Education, Hudson, NY.

SEPUP (Science Education for Public Understanding)
Middle and Secondary School, 2-year long courses.
Developed by Lawrence Hall of Science Berkeley CA, 1995.
Available: Lab-Aids, Ronkonkoma, NY.

STC (Science and Technology for Children)
K-6 Curriculum, 24 units, inquiry-based and dealing with technological applications of science.
Developed by National Science Resources Center, 1991
Available: Carolina Biological Supply, Burlington, NC.

Appendix B Activity Books and Projects

Abruscato, Joseph.(2000). *Whizbangers and Wonderments: Science Activities for Young People.* Boston: Allynn and Bacon.

AIMS (Activities for Integrating Mathematics and Science). AIMS Education Foundation, PO Box 8120, Fresno, CA 93747-8120.

American Institute of Physics. (1999). *The Best of Wonder Science.* New York: Delmar Publishers.

Baeckler, Virgina. (1986). *Storytime Science.* 26 Hart Avenue, Hopewell, NJ: Sources.

The Best of Wonder Science: Elementary Science Activities. (1997). Albany, NY: Delmar Publishers.

Devito, A & Krockover, G.H. (1991). *Creative Sciencing: Ideas and Activities for Teachers and Children*, Glenview, IL: Scott Foresman (Good Year).

The Earth Works Group. (1990). *50 Simple Things Kids Can Do to Save the Earth.* New York: Andrews & McMeel.

FACETS (Foundations and Challenges to Encourage Technology-based Science), American Chemical Society, Dubuque: Iowa, Kendall/Hunt.

GEMS (Great Explorations in Math and Science). University of California, GEMS, Lawrence Hall of Science, Berkeley, CA 94720-5200.

Hassard. Jack. (1990). *Science Experiences: Cooperative Leaning and the Teaching of Science.* New York: Addison- Wesley.

Hoehn, Robert (1993). *Science Starters: Over 1000 Ready-to-Use Attention Grabbers that Make Science Fun for grades 6-12.* West Nyack, NY: The Center for Applied Research in Education.

Kane, Patricia et.al. (1992). *Bridges to the Natural World.* New Jersey Audubon Society 790 Ewing Avenue. PO Box 125, Franklin Lakes, NJ 07417-2271.

Lowery, L.F., & Verbeeck, C. (1987). *Explorations* (3 volumes: earth science, physical science, life science). Carthage, Il: Fearon.

Project Wet. (1995). *Curriculum and Activity Guide.* Houston, Texas: Council for Environmental Education.

Project Wild. (1994). *Activity Guide*. Western Regional Environmental Council.

Roa, Michael. (1993). *Environmental Science Activities Kit*. West Nyack, NY: The Center for Applied Research in Education.

Sisson, Edith. (1982). *Nature with Children of All Ages*. Massachusetts Audubon Society. New York: Prentice Hall.

Strongin, H. (1991). *Science on a Shoestring*. Reading, MA: Addison-Wesley.

Van Cleave, J. P. (1989-?). *Science for Every Kid* (6 volumes: biology, chemistry, earth, astronomy, physics, and geography.

NEW JERSEY

Activities in Science. The Center for Elementary Science, Fairleigh Dickinson University.

Grant, David. *The Topic is Sandy Hook*. Sandy Hook, NJ: Ocean Institute, Brookdale Community College, Box 533, Sandy Hook NJ 07732. (08) 872-2284.

New Jersey Department of Environmental Protection. (1997) *Beneath the Shell*. Office of Environmental Planning CN 418 Trenton, NJ 08625. (609) 292-2113. Free.

Appendix C Organizations

American Chemical Society
1155 16th St., N.W.
Washington, D.C. 20036
(202) 872-4600; (800) 227-5558
http://www.acs.org
http://www.chemcenter.org

American Forest Foundation
1111 19th St., N.W., Suite 780
Washington, DC 20036
(202) 463-2462
http://www.affouondation.org

American Geological Institute
4220 King St.
Alexandria, VA 22301-1507
(703) 379-2480

American Meteorological Society
1200 New York Ave., N.W., Suite 410
Washington, DC 20006
(202) 466-5728

American Physical Society
American Center for Physics
One Physics Ellipse
College Park, MD 20740
(301) 209-3200
http://aaps.org.educ

ASCD (Association for Supervision and Curriculum Development)
1250 N. Pitt St.
Alexandria, VA 22314-1403
(703) 549-9110
http://www.ascd.org

Cousteau Society
870 Greenbrier Circle, Suite 402
Chesapeake, VA 23320-9864
(757) 523-2747; Fax: (757) 523-2747

Council for Exceptional Children
1920 Association Dr.
Reston, VA 20101–1589
(703) 620-3660; (703) 264-9446 (TTY)
http:www.cec.sped.org

Ecological Society of America
2010 Massachusetts Ave., N.W. , Suite 400
Washington, DC 20036
(202) 833-8773

(ENC)Eisenhower National Clearinghouse for Mathematics and Science Education
The Ohio State University
1929 Kenny Rd.
Columbus, OH 43210-1079
(614) 292-7784; (800) 621 –5785
http://www.enc.org

Geological Society of America
3300 Penrose Pl.
P.O. Box 9100
Boulder, CO 80301-9140
(303) 447-2020; (800) 472- 1988
http://www.geosociety.org

International Society for Technology Education
University of Oregon
1787 Agate St.
Eugene, OR 97403-1923
(541) 346-4414; (800) 336-5191
http:www.iste.org

NASA(National Aeronautics and Space Administration),(Education Division
Code FE, NASA Headquarters
300 E St. S.W.
Washington, DC 20546
(202) 358-1110
http:www.hq.nasa.gov/office/codef/education

(NARST)National Association for Research in Science Teaching
c/o Arthur L. White
NARST Executive Secretary
1929 Kenney Rd.,Rm 2200E
The Ohio State University
Columbus, Ohio 43210
(614) 292-3339
http://science.coe.uwf.edu/narst/narst.html

National Association of Biology Teachers
1320 N Street, N.W.
Washington, DC 2005

National Association of Geoscience Teachers
Department of Geology-9080
Western Washington University
Bellingham, WA 98225-9080
(360) 650-3587

National Audubon Society
700 Broadway
New York, NY 10003
(212) 979-3000

National Center for Improving Science Education
2000L St., N.W.,Suite 603
Washington, DC
(202) 467-0562
http://www.wested.org

National Earth Science Teachers Association
2000 Florida Ave., N.W.
Washington, DC 20009
(202) 462-6910; (800) 966-2481

National Energy Foundation
5225 Wiley Post Way, Suite 170
Salt Lake City, UT 84116
(801) 539-1406
http://www.nefl.org

National Gardening Association
180 Flynn Ave.
Burlington, VT 05401
(802) 863-1308; (800) 538-7476

http://www.garden.org

National Geographic Society
1145 17th St., N.W.
Washington, D.C 20036
(202) 857-7000; (800) 368-3728
http://www.nationalgeographic.com

National Marine Educators Association
P.O. Box 1470
Ocean Springs, MS 39566-1470
(601) 374-7557
http://www.marine-ed.org

National Middle Level Science Teachers Association
C/o Rowena Hubler
Ohio Department of Education
65 S. Front St.
Columbus, OH 43215-4183
(614) 466-2761
http://www.nsssta.org/nmlsta/index.htm

NSELA (National Science Education Leadership Association)
P.O. Box 5556
Arlington, VA 22205
(703) 524-8646

NSTA (National Science Teachers Association)
1840 Wilson Blvd.
Arlington, VA 22201-3000
(703) 243 –7100
http://222.nsta.org

National Weather Association
6704 Wolke Ct.
Montgomery, AL 36116-2134
(334) 2213-0388

National Wildlife Federation
89225 Leesburg Pike
Vienna, VA 221184-0001
(703) 790-40000
htttp://.nwf.org

North American Association for Environmental Education
1255 23rd St/.N.W., Suite 400
Washington, DC 20037-1199
(202) 884-8912
http://eelink.umich.edu/naaee.html

Project WET, The Watercourse Program
201 Culbertson Hall
Montana State University
Bozeman, MT 59717-5392
(406) 994-5392
http://www.montana.edu/wwwwet

Project WILD
5430 Grosvenor Lane, Suite 230
Bethesdaa, MD 20814
(301) 493-5447
http://eelink.umich.edu.wild

Raptor Education Foundation
21901 E. Hampden Ave.
Aurora, CO 80013
(303) 680-8500

School Science and Mathematics Association
Department of Curriculum and Foundations
Bloomsburg University
499 E. Second St.
Bloomsburg, PA 17815-1301
(717) 389-4915
http://hubble.bloomu.edu/~ssma

Science-by-Mail
Museum of Science
Science Park
Boston, MA 02114-1099
(617) 589-0437; (800) 729-3300
http://www.mos,org/mos/sbm/sciencemail.html

Smithsonian Institution, Office of Education
MRC 402
Arts and Industries Bldg, Rm 1163
Washington, DC 20560
(202) 357-2425
http://educate.si.edu

Soil and Water Conservation Society
7517N.E., Ankeny Rd.
Ankeny, IA 50021
(515) 289-2331; (800)THE SOIL
http://www.swcs.org

CSREES (US Department of Agriculture, Cooperative State, Research, Education and Extension Services)
Rm. 3441, South Building
Washington, DC 2050
(202) 720-58523

NOAA (US Department of Commerce, National Oceanic and Atmospheric Administration)
NOAA Public Affairs Correspondence Unit
1305 East-West Hwy., Stn 1W204
Silver Spring, MD 20910
(301) 713-1208

US Department of Education, Office of Educational Research and Improvement
555 New Jersey Ave., N.W.
Washington, DC 20808-5645
(202) 219-2116
http://www.ed.gov/offices/OERI/oeribro.html

US Space Foundation
2860 S. Circle Dr., Suite 2301
Colorado Springs, CO 80906-4184
(719) 576-8000; (800) 691-4000
http://www.usaf.org

World Wildlife Fund
1250 24th St., N.W.
Washington, DC 20037-1175
(202) 293 –4800
http://www. wwf.org

Appendix D Periodicals and Newsletters

For Teachers

Note: All addresses are available in Appendix C, Organizations or Appendix E, Suppliers)

AIMS Magazine (Teachers of grades 3-8, Science and math activities, 10 per year, $30, AIMS Education Foundation)

Air & Space/Smithsonian (Articles for teachers on history and technology of aerospace, 6 per year, $20, Air & Space/Smithsonian)

Appraisal Science Books for Young People (Teachers of grades PreK –8, Reviews science trade books written for students, 4 per year, $45, Appraisal)

Audubon (Articles for teachers on the environment, protection of wildlife, and wildlife, 6 per year, $20, National Audubon Society)

Connect (Teachers of grades K-8, articles supporting hands-on learning in science, and mathematics, 5 per year, $20, Teacher's Laboratory)

Discover: The World of Science (Easy to understand articles for teachers about all areas of science, 12 per year, $24.95, Discover)

International Wildlife (Articles about wildlife, 6 per year $16, national Wildlife Federation)

Journal of Research in Science Teaching (Research in science education, 10 per year, $450 or $77 for NARST members, NARST)

National Wildlife (Articles on wildlife and the environment, 6 per year, 16, National Wildlife Federation.)

Natural History (Articles published by American Museum of Natural History on all related areas of science, 12 per year, $30, Natural History)

Science (Easy to read information about scientific events, 10 per year, AAAS)

Science Activities (Activities for teachers of grades 1-8, 4 per year, 32.00, Heldref)

Science and Children (Articles for teachers of grades K-8 emphasizing hands-on activities, 8 per year, $55, NSTA)

Science Books and Films (Critical reviews of books and electronic resources in science mathematics and technology, 9 per year, $40, Science Books)

Science is Elementary (Resource magazine for teachers of K-6, Museum Institute for Teaching Science)

Science News (General information on scientific events, weekly, $49.50 per year, Science Service)

Science Scope (Articles for teachers of Middle School emphasizing hands-on activities, 8 per year, $55, NSTA)

School Science and Mathematics

Smithsonian (Articles about science, humanities, and arts, 12 per year, $24, Smithsonian Institute)

WETnet. Newsletter (Project WET)

Special Interest or Upper Grades

American Biology Teacher (Articles for teaching, aimed mainly at middle school and high school teachers, 9 per year, National Association of Biology Teachers)

*Journal of Chemical Education (*Articles on chemistry, at high school and college level, 12 per year, $34, Journal of Chemical Education)

Journal of Geological Education (6 per year, National Association of Geology Teachers)

Oceanus (Report on research at the Woods Hole Oceanographic Institute, 2 per year $15, WHOI Publication Service)

Physics Teacher *(9 per year, American Association of Physics Teachers)*

Science Teacher (Articles for teachers of Middle School and High School emphasizing current developments and innovations in science teaching, 9 per year, $56, NSTA)

The Science Times (Articles about science in the news geared mainly for teachers of upper elementary through high school, 10 per year, $50)

For Students

Note: All addresses are available in Appendix C, Organizations or Appendix E, Suppliers)

3-2-1 Contact (Grades 3-8, 3-2-1 Contact Magazine10 per year, $19.90 per year)

Dolphin Log (Grades 2 –8; Cousteau Society, 6 per year, $15 per year)

Dragonfly (Grades 3-6; NSTA, 5 per year, prices vary according to number purchased, $6 per year for class sets of 20 or more with one Teacher's companion issue)

Kids Discover (Grades 1-8, Kids Discover, 10 per year, $19.95 per year)

Muse (Grades 1-8, Muse, 6 per year, $24 per year)

National Geographic World (Grades 3-8, National Geographic Society, 12 per year, $17.95 per year)

Odyssey (Grades 4-8, Cobblestone, 9 per year, 26.95 per year)

Ranger Rick (Grades 1-6, National Wildlife Federation, 12 per year, $15 per year)
Preschool and primary grades *Your Big Backyard,* 10 per year)

Science Weekly (Grades 1-6, Science Weekly, 20 per year)

Science World (Grades 6-8, Scholastic,18 per year, $7.50 per year with a minimum order of 10)

WonderScience (for grades 4-6) American Chemical Society)

Appendix E –Suppliers

3-2-1 Contact
P.O. Box 51177
Boulder, CO 80322-1177
(800) 678-0613

Air & Space /Smithsonian
P.O. Box 420113
Palm Coast, FL 32142-0113
(800) 766-2149

Appraisal
605 Commonwealth Ave.
Boston, MA 02215

AIMS Education Foundation
P.O. Box 8120
Fresno, CA 93747-8120
(209) 255 –4094
Fax: (209) 255-6396

Carolina Biological Supply Co.
2700 York Rd.
Burlington, NC 27425-5665
Fax: (800) 535 -2669

CESI (Council for Elementary Science)
C/0 John Penick
789 Van Allen Hall
University of Iowa
Iowa City, IA 52242
(319) 335-1183
Fax (319) 335-1188

Cobblestone Publishing
7 School St.
Peterborough, NH 03458-1454
(800) 924-7209
Fax (603) 924-7380

Delta Education
P.O. Box 915
Hudson, NH 03051 (800) 258-1302

Discover
P.O. Box 420105
Palm Coast Fl 32142-0105

Edmund Scientific
102 R. Gloucester Pike
Barrington, NJ

Flinn Scientific
P.O. Box 219
Batavia, IL 60510
(630) 879-6900
Fax: (630) 879-6962
e-mail:flinnsci@aol.com

Frey Scientific Company
905 Hickory lane
Mansfield, OH 44905

Heldref Publications
1319 18th St. N.W.
Washington, DC 20036-1802
(800) 365 –9753; (202)296-6267
Fax: (202) 296-5149

Journal of Chemical Education
Subscription and Book Order Department
1991 Northampton St.
Easton, PA 18042
(800) 991-5534; (608) 262-7146
Fax: (608) 262-7145
e-mail:jce@aol.com
http://jchemed.chem.wisc.edu

Kendall/Hunt Publishing Company
4050 Westmark Dr.
Dubuque, IA 52002
(800) 772-9165
http://www.kendallhunt.com

Kids Discover
P.O. Box 5205
Boulder, CO 80322 (800) 284-8276

Lab-Aids
17 Colt Ct.
Ronkonkoma, NY 11779
(516) 737- 1286

Lawrence Hall of Science
University of California
Berkeley, CA 94720-5200
(510) 642-7771
Fax: (510) 643-0309

Lego Dacta
555 Taylor Rd.
P.O. Box 1600
Enfield, CT 06083-1600
(800) 527-8339
Fax: (203) 763-2466

Muse
Box 7468
Red Oak, IA 51591-2468
(800) 827-0227

Natural History
American Museum of Natural History
Central Park West at 79th St.
New York, NY 10024
(212) 769-5304
Fax: (212) 296-7119

NASCO
901 Janesville Ave.
P.O. Box 901
Fort Atkinson, WI 53538-0901
(800) 558 –9595; (414) 563-2446
Fax: (414) 563-8296

Ohaus Scale Corporation
29 Hanover Road
Florham Park, NJ 07932

Sargent-Welch/VWR Scientific
911 Commerce Ct.
P.O. Box 5229
Buffalo Grove, IL 60089-5229
(800) 727 –4368
Fax: (800) 676-2540
e-mail:sarwel@sargentwelch.com
http://www. Saaargentwelch.com

Scholastic (To order Science World)
P.O. Box 3710
2931 E. McCarty St.
Jefferson City, MO 65102-3710
(800) 631-1586

Science Books & Films
Department SBF
P.O. Box 3000
Denville, NJ 07834
(202) 326 –6454
Fax: (202) 371-9849

Science Kit, Inc.
777 E. Park Drive
Tonawanda, NY 14150

Science Service
Subscription Department
P.O. Box 1925
Marion, OH 43305
(800) 247-2160
Fax: (614) 382-5866
e-mail:scinews@scisvc.org
http://sciencenews.org

Science Weekly
P.O. Box 70154
Washington, DC 20088

Smithsonian Institution
(To order *Smithsonian* magazine)
Membership Data Center
P.O. Box 42039
Palm Coast, FL 32142-9143
(800) 766-2149

Teacher's Laboratory
P.O. Box 6480
Brattleboro, VT 05302-6480
(800) 769-6199
Fax: (802) 254-5233
e-mail: connect@sover.net

Wards Natural Science Establishment
P.O. Box 92912
Rochester, NY 14692-9012
(800) 962-2660
Fax: (800) 635-8439

WHOI Publication Services
P.O. Box 50145
New Bedford, MA 02745
(508) 457-2000
Fax: (508) 457-2180
e-mail: oeanusmag@whoi.edu
http://www.whoi.edu.oceanus

Appendix F Technology

Organizations, Special Resources, Periodicals and Suppliers of Software and Educational Technology

Enter

Periodical for upper elementary school focusing on computer games and ideas.
One Disk Drive
P.O. Box 2686
Boulder, CO, 80322
10 issues per year

EPIE Institute
103-3 W. Montauk Hwy.
Hampton Bays, NY 11947
(516) 728 –9100
http://www.epie.org
Organization evaluates educational products provides programs to help communities up-date programs technologically, publishes materials with information on programs

IDEAAAS: Sourcebook for Science Mathematics and Technology Education

Washington DC: American Association for the Advancement of Science
Armonk, NY: The Learning Team, 1995.

International Society for Technology Education
University of Oregon
1787 Agate St.
Eugene, OR 97403-1923
(541) 346-4414; (800) 336-5191
http:www.iste.org
Organization focuses on improving education, using technology. It provides programs, workshops and produces publications.

Learning and Leading with Technology

Periodical focuses on technology in the classroom for use with grades K-8. Available from International Society for Technology in Education
(8 per year, $65 per year)

Queue, Inc.
338 Commerce Drive
Fairfield, CT 06432
(800) 232-2224 (203) 335-0906
Educational C- ROMS and Software

Scholastic Software & Multimedia
2931 East McCary St.
Jefferson City, MO 65101
(800) 724-6527

Sunburst Educational Software Dept.
EG53, 101 Castleton St.
PO Box 100
Pleasantville, NY 10570
(800) 321-7511
http://www.nysunburst.com

Teacher's Video Company
Division of Global Video, LLC
P.O. 4455-02SC01
Scottsdale, Arizona 85261
Science Videos

TERC
2067 Massachusetts Ave.
Cambridge, MA 02140
(617) 547-0430
http://www.terc/edu
Organization focuses on science and mathematics teaching. It provides several electronic community services, produces software for data collection, and publishes a periodical on science, mathematics and technology education.

Tom Snyder Productions80 Coolidge Hill Road
Watertown, MA 02472-5003
(800) 324-0236
Fax (617) 926-6222
www.teachtsp.com
CD-ROMS, Discs, Interactive Group Software

Bibliography

Abruscato, Joseph. (2000). *Teaching Children Science.* Boston: Allynn and Bacon.

Armstong, Thomas. (1994). *Multiple Intelligences in the Classroom..* Alexandria, VA: ASCD.

Association for Supervision and Curriculum Development. (1997). *Only the Best: The Annual Guide to the Highest-Rated Educational Software and Multimedia* 1997-1998. Alexandria Va.: ASCD.

Ausubel, D.P. (1963). *The Psychology of Meaningful Verbal Learning.* New York: Grune and Stratton.

Bentley, Michael et al (2000). *The Natural Investigator.* Belmont, CA: Wadsworth.

Bloom, B.S. (1956). *Taxonomy of Educational Objectives. Handbook I. Cognitive Domain.* New York: David McKay.

Bruner, J. S. (1963). *The Process of Education.* New York: Vintage Books.

Carin, Arthur C. (1997). *Guided Discovery Activities for Elementary School Science.* Upper Saddle River, NJ: Prentice Hall.

Carin, Arthur C. (1997). *Teaching Modern Science.* Upper Saddle River, NJ: Prentice Hall.

Chiapetta, Eugene L. et al (1998). *Science Instruction in the Middle and Secondary Schools.* Upper Saddle River, NJ: Merrill.

Dobey, Daniel C. et.al. (1999). *Essentials of Elementary Science.* Boston: Allynn and Bacon.

Dewey, John. (1916, Paperback ed., 1961). *Democracy and Education.* New York: Macmillan.

Ebenezer, Jazlin and Eddy Lau. (1999). *Science on the Internet.* Upper Saddle River, NJ: Prentice Hall.

Educators Guide to Free Science Materials. (1997). Randolph, Wisconsin: Educators Progress Service Inc.

Esler, William K. and Mary K. (1996). *Teaching Elementary Science.* Boston: Wadsworth Publishing Co.

Farmer, Walter A. et al (1991). *Secondary Science Instruction: An Integrated Approach.* Providence, R.I.: Janson Publications.

Flinn Scientific Staff. (2000). *School Science and Safety.* Batavia, Illinois: Flinn Scientific.

Foster, Gerald. Elementary Mathematics and Science Methods: Inquiry Teaching and Learning. (1999). New York: Wadsworth Publishing Co.

Gega, Peter C.and Joseph Peters. (1998). *Concepts and Experiences in Elementary School Science.* Upper Saddle River, NJ: Merrill.

Gega, Peter C.and Joseph Peters. (1998) *How to Teach Elementary Science.* Upper Saddle River, NJ: Merrill

Harlan, J. and Rivkin, M (2000). *Science Experiences for the Early Childhood Years: An Integrated Approach.* Upper Saddle River, NJ: Merrill, Prentice Hall

Hein, George and Sabra Price. (1994). Active Assessment for Active Science. Portsmouth,NH: Heinemann.

Howe, Ann C and Linda Jones. (1998). *Engaging Children in Science.* Upper Saddle River, NJ: Prentice-Hall.

Hunter, Madeline. (1986). *Mastery Teaching.* El Segundo, CA: TIPS Publications.

The International Assesment of Educational Progress. (1992). *Learning Science.* Princeton, New Jersey: Educational Testing Service.

Johnson, D.W. and R. Johnson. (1975). *Learning Together and Alone.* Englewood Cliffs, NJ: Prentice-Hall.

Kellough, Richard et al (1996). *Integrating Mathematics and Science.* Englewood Cliffs, NJ: Prentice Hall.

Krajcik, Joseph et al (1999). *Teaching Children Science: A Project-Based Approach.* New York: McGraw Hill.

Kuhn, Thomas. (1970). *The Structure of Scientific Revolutions*, Second ed. Chicago: University of Chicago Press.

Lapointe, Archie E. et al. (1989). *A World of Differences: An International Assessment of Mathematics and Science.* Princeton, New Jersey: Educational Testing Service.

Martin, Ralph et al (1998). *Science For All Children.* Boston: Allynn and Bacon.

National Research Council. (1990). *Fulfilling the Promise: Biology Education in the Nation's Schools.* Washington, D.C.: National Academy Press.

National Research Council. (1996). *National Science Education Standards.* Washington, DC: National Academy Press.

National Science Resource Center. (1996). *Resources For Teaching Elementary School Science.* Washington, DC: National Academy Press.

National Science Resource Center. (1998). *Resources For Teaching Middle School Science.* Washington, DC: National Academy Press.

Phelan, Carolyn. (1996). *Science Books for Young People.* Chicago, ILL.: American Library Association Booklists Publications.

Piaget, Jean and Barbel Inhelder. (1969) *The Psychology of the Child.* New York, NY: Basic Books.

Rezba, Richard. (1996). *Readings for Teaching Science in Elementary and Middle Schools.* Dubuque, IA:Kendall/Hunt Publishing Company.

Rezba, Richard et al (1995). *Learning and Assessing Science Process Skills..* Dubuque, IA:Kendall/Hunt Publishing Company.

Schmidt, William et al (1997). *A Splintered Vision: An Investigation of US Science and Mathematics Education.* Norwell, MA:Kluwer Academic Publishers.

Sherman, Sharon. (2000). *Science and Science Teaching.* New York: Houghton Mifflin.

Slavin, R. (1983). *Cooperative Learning.* New York: Longman.

TIMSS. (1997). *Third International Mathematics and Science Study.* Philadelphia, PA:
Mid-Atlantic Eisenhower Consortium for Mathematics and Science Education.

Trowbridge, Leslie W. and Bybee, Roger W. (1990). *Becoming A Secondary Science Teacher.* Columbus, Ohio: Merrill Publishing Company.

Victor, Edward. And Richard D. Kellough (2000). *Science for the Elementary and Middle School.* Upper Saddle River, NJ: Merrill.

Vygptsky, L. (1978). *Mind in Society.* Cambridge, MA: Harvard University Press.

NEW JERSEY

New Jersey Department of Education. (1998). *New Jersey Science Curriculum Framework.* Trenton, NJ: Department of Education.

New Jersey Department of Education (1996). *Core Curriculum* Content Standards. Trenton, NJ: Department of Education.

Index

Author Biographical Sketch

Dr. Sharon Brendzel is a Full Professor at Kean University. Her teaching includes methodology courses in science education. She is also the coordinator of the Liberal Studies in Mathematics, Science and Technology major for elementary education students and the coordinator of the Transition to Teaching Grant funded by the US Department of Education. Dr. Brendzel is Co-PI for two National Science Foundation grants, the NJ Math Science Partnership and the Noyce Scholars program at Kean University. She has also published a large number of articles in national magazines including several in *Science Scope*. She is a frequent presenter at national, regional and local conferences.